Green Energy and Technology

For further volumes:
http://www.springer.com/series/8059

Silvia Daniela Romano · Patricio Aníbal Sorichetti

Dielectric Spectroscopy in Biodiesel Production and Characterization

Springer

Prof. Dr. Eng. Silvia Daniela Romano
Grupo de Energías Renovables
Facultad de Ingeniería
Universidad de Buenos Aires
Av. Paseo Colón 850
1063 Buenos Aires
Argentina
e-mail: sromano@fi.uba.ar

Prof. Eng. Patricio Aníbal Sorichetti
Laboratorio de Sistemas Líquidos
Facultad de Ingeniería
Universidad de Buenos Aires
Av. Paseo Colón 850
1063 Buenos Aires
Argentina
e-mail: psorich@fi.uba.ar

and

Consejo Nacional de Investigaciones
Científicas y Técnicas (CONICET)
Av. Rivadavia 1917
1033 Buenos Aires
Argentina

ISSN 1865-3529

ISSN 1865-3537 (eBook)

ISBN 978-1-4471-5872-1

ISBN 978-1-84996-519-4 (eBook)

DOI 10.1007/978-1-84996-519-4

Springer London Dordrecht Heidelberg New York

British Library Cataloguing in Publication Data
A catalogue record for this book is available from the British Library

Cover design: eStudio Calamar, Berlin/Figueres

Printed on acid-free paper

Springer is part of Springer Science+Business Media (www.springer.com)

Contents

Chapter 1
Introduction

1.1 Introduction to Biofuels

At the beginning of the twenty-first century, most liquid and gaseous fuels used in the world are obtained from oil and natural gas: compressed natural gas (CNG), liquefied petroleum gas (LPG), gasoline (petrol), kerosene, diesel-oil, fuel-oil, aviation fuel, etc. However, petroleum and natural gas are non-renewable energy sources, since they were originated by geological processes during millions of years, enormously slower than their present consumption rate. Although new oil and gas fields are still being discovered, it is clear that easily accessible reserves will run out in a few decades, and many countries are already facing serious energy shortages.

In the last decade of the twentieth century, especially after 1995, the scientific community became seriously concerned about the increase of global temperatures due to the steady increase, since the beginning of the Industrial Revolution, of the atmospheric concentration of the so-called "greenhouse gases", particularly carbon dioxide (CO_2) originated in the use of fossil fuels. Increasing public awareness on the serious consequences for the environment led to the signature of the Kyoto Protocol in 1997, requiring that 35 industrialized countries reduce their emissions of greenhouse gases by an average of 5.2% between 2008 and 2012, in comparison to 1990 levels [1]. Upon ratification by 55 parties, the treaty came into force in November 2004. In December 2009, representatives of most countries in the world met in Copenhagen at the COP 15 [2, 3], Conference organized by United Nations, to discuss the actions to be taken for the next years, related to the climate change specially during the period between 2012 and 2020.

Widespread concern about scarcity of supplies and adverse environmental effects, together with the instabilities of international oil prices and the difficulty of providing energy to many isolated areas in developing countries, are some of the reasons that make necessary to promote the prompt development and utilization of

S. D. Romano and P. A. Sorichetti, *Dielectric Spectroscopy in Biodiesel Production and Characterization*, Green Energy and Technology, DOI: 10.1007/978-1-84996-519-4_1, © Springer-Verlag London Limited 2011

renewable and environmentally friendly energy resources. These include biofuels, wind, solar, tidal and geothermal.

As distinct from fossil fuels, biofuels are produced from biomass. In this context, biomass may be defined as biological material derived from living, or recently living, organisms and from which it is technologically viable to obtain energy. Forestry and agricultural residues, food waste and energy crops are the most important present sources of biomass. Biofuels have several advantages, making them an attractive option to diversify the energy supply matrix: they are obtained from renewable sources, generate fewer emissions, and are biodegradable. Moreover, several biofuels can be produced from non-food residues and substances.

A straightforward classification of biofuels is based on their state (at room temperature); gaseous biofuels, like biogas from different sources and syngas (coal gas); liquid biofuels including biodiesel, bioethanol, vegetable oil and bio-oil; and solid biofuels such as wood (used in different forms by humankind since prehistoric times), biomass briquettes, sawdust and charcoal.

Each biofuel is obtained from specific raw materials and production processes (physical, chemical, biochemical, etc.) and in consequence they have different uses:

- *Biodiesel* is commonly produced by the chemical reaction (transesterification) of the vegetable oil or animal fat feedstock with a short-chain aliphatic alcohol (typically methanol or ethanol). It has the same uses as diesel-oil and it is often blended with it.
- *Bioethanol* is obtained from the aerobic fermentation of carbohydrate-rich biomass, and is a replacement for gasoline (petrol) often used in blends.
- *Bio-oil* is extracted by pyrolysis from forestry or agricultural residues. It can be used in boilers and furnaces.
- *Biogas* results from the anaerobic fermentation of certain animal or urban residues, and replaces natural gas.
- *Syngas* is generated from biomass by a gasification process. It has a lower heating value than natural gas, and it can be used in internal combustion engines and for heating purposes.
- *Charcoal* is produced by the slow pyrolysis of biomass, usually wood, by heating in the absence of oxygen. It replaces coal in many developing countries.
- *Biomass briquettes* are made from agricultural wastes, sawdust or wood, among other substances, by a slow pyrolysis process followed by compression. In many developing countries are a convenient alternative to wood for cooking and heating applications, and also can be used in industrial boilers.

Given the many kinds of biofuels presently available, the reader could question why biodiesel and bioethanol are the most widely known and used worldwide. The foremost reason is that these biomass-derived fuels are mainly used in automotive applications, blended with diesel-oil and gasoline respectively. However, biodiesel utilization is not restricted to land transportation, and in several countries it has been successfully used in military and commercial jet aircraft (for instance, in blends with JP1 aviation fuel) and also in ships.

It must be borne in mind that large-scale production and utilization of biodiesel is conditioned, not only by technological factors, but also by economical and political issues in each country. The European Community has been producing and using biodiesel for several years. Germany is the leading world producer, using rapeseed oil as raw material. Other important European producers are France and Italy. Large quantities of biodiesel are also produced in the United States, from soybean oil, and several Asian countries export biodiesel from palm oil. In Central and South America most countries are starting to enact legislation on biofuels, some of them, such as Argentina and Brazil, are producers for some time now. These two countries have the greatest potential for biofuel production in South America, taking into account the availability of land, climate and soil diversity.

From the previous comments, it is easy to understand that an international certification system would be very convenient, both for producers and users of biofuels. Such certification systems should address a variety of issues related to not only to greenhouse gas emissions and energy consumption, but also to other equally important dimensions of sustainability: ecological (land use and protection of natural biospheres) and social (workers health and safety, labor rights, land rights, etc.). At present, several international entities are developing schemes along these lines.

The most frequently voiced criticism against biodiesel is that it is detrimental to food supplies, since it depends from the same foodstuffs. However, second generation biodiesel can be produced using non-edible oils, such as jatropha, cotton or tung. Research is under way to use oil from other species that thrive in marginal agricultural land, where conventional crops will produce low yields, and also from micro-algae (which can produce a much larger volume of oil per unit area than land crops) or even fungus. It must also be remarked that biodiesel can be obtained from waste vegetable oil (thus recycling a contaminant waste) and also from animal fats, a residue from meat packing plants that is available at a very low cost.

1.2 Production of Biodiesel in the European Union

In Fig. 1.1, it is shown the growth of the biodiesel production in the European Union during the last years [4]. It can be seen that the maximum increments were between 2004 and 2006 (approximately 65% from 2004 to 2005 and 54% from 2005 to 2006).

Figure 1.2 shows the evolution of the production capacity of biodiesel in the European Union during the last years [4]. It can be seen that since 2007 the production of biodiesel in the European Union was approximately 50% of its production capacity. The main reason is the scarcity of raw material [5].

In Fig. 1.1, it is shown the growth of the biodiesel production in the European Union during the last years [4]. It can be seen that the maximum increments were

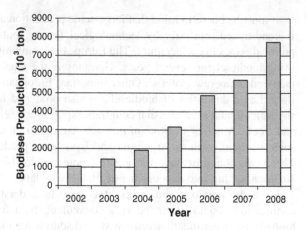

Fig. 1.1 Production of bio-diesel in the European Union during the last years

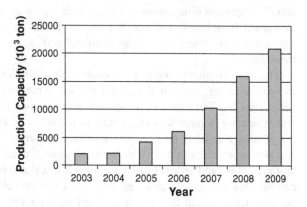

Fig. 1.2 Production capacity of biodiesel in the European Union during the last years

between 2004 and 2006 (approximately 65% from 2004 to 2005 and 54% from 2005 to 2006).

1.3 Production of Biodiesel in the World

During the last years the production of biodiesel in the world has increased rapidly. Data for the leading producers of biodiesel during 2007 are plotted in Fig. 1.3, whereas in Fig. 1.4 the data correspond to 2008 [4, 5], [6]. In both figures, the volume of production of each country is represented.

From the comparison of both figures it is clear that:

- Germany, the United States of America and France were the three most important producers.
- The productions of Germany and Malaysia have remained practically constant.

Fig. 1.3 Leading producers
of biodiesel in the world
during 2007

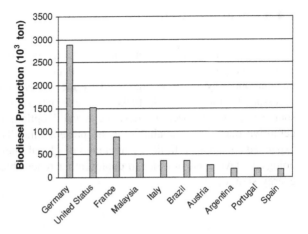

Fig. 1.4 Leading producers
of biodiesel in the world
during 2008

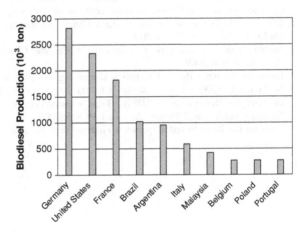

- The production of USA and France increased more than 50 and 100%,
  respectively.
- Brazil and Argentina increased their production in almost 200% and more than
  400%, respectively.
- Belgium and Poland have been incorporated to the "top ten".

Other important producers in 2008 were: Denmark and Sweden, Austria, Spain,
United Kingdom, Slovakia, Greece, Hungary, Czech Republic, Indonesia, etc.

Estimations for year 2009 [5] indicate that the first five producers of 2008 will
remain the same but some of them in different places. Germany and Argentina
probably will remain in the same positions [5].

1.4 Concluding Remarks

The mitigation of adverse environmental effects and the diversification of the energy supply matrix (particularly in "oil-dependent" countries) are two concerns that undoubtedly must be addressed by the present generation. Therefore, it is necessary to find renewable fuels as an alternative to those obtained from fossil sources; in this respect biofuels are a valid option, certainly not the only or definitive one, that is available at the present time. Their use can be gradually expanded to satisfy a limited percentage of global energy needs, while the technology is developed in order to deploy other renewable and economically viable energy sources.

References

1. United Nations (1998) Kyoto Protocol to the United Nations Framework Convention on Climate Change. http://unfccc.int/resource/docs/convkp/kpeng.pdf. Accessed 6 Nov 2009
2. COP 15 (2009) United Nations Climate Change Conference Copenhagen 2009. http://en.cop15.dk/. Accessed 30 Nov 2009
3. United Nations Framework Convention on Climate Change. http://unfccc.int/2860.php. Accessed 6 Nov 2009
4. European Biodiesel Board. http://www.ebb-eu.org/stats.php. Accessed 30 Nov 2009
5. St James C (2009) Estado de la Industria Argentina de Biodiesel. Cámara Argentina de Energías Renovables (CADER). http://www.argentinarenovables.org/archivos/Biodiesel SegundoCuatrimestre2009.pdf. Accessed 26 Nov 2009
6. National Biodiesel Board. http://www.biodiesel.org/. Accessed 6 Nov 2009

Chapter 2
Introduction to Biodiesel Production

2.1 Introduction

Biodiesel [1–5] is a liquid biofuel obtained by chemical processes from vegetable oils or animal fats and an alcohol that can be used in diesel engines, alone or blended with diesel oil.

ASTM International (originally known as the American Society for Testing and Materials) defines biodiesel as a mixture of long-chain monoalkylic esters from fatty acids obtained from renewable resources, to be used in diesel engines.

Blends with diesel fuel are indicated as "Bx", where "x" is the percentage of biodiesel in the blend. For instance, "B5" indicates a blend with 5% biodiesel and 95% diesel fuel; in consequence, B100 indicates pure biodiesel.

2.1.1 Advantages of the Use of Biodiesel

Some of the advantages of using biodiesel as a replacement for diesel fuel are [1–4]:

- Renewable fuel, obtained from vegetable oils or animal fats.
- Low toxicity, in comparison with diesel fuel.
- Degrades more rapidly than diesel fuel, minimizing the environmental consequences of biofuel spills.
- Lower emissions of contaminants: carbon monoxide, particulate matter, polycyclic aromatic hydrocarbons, aldehydes.
- Lower health risk, due to reduced emissions of carcinogenic substances.
- No sulfur dioxide (SO_2) emissions.
- Higher flash point (100°C minimum).

S. D. Romano and P. A. Sorichetti, *Dielectric Spectroscopy in Biodiesel Production and Characterization*, Green Energy and Technology, DOI: 10.1007/978-1-84996-519-4_2, © Springer-Verlag London Limited 2011

- May be blended with diesel fuel at any proportion; both fuels may be mixed during the fuel supply to vehicles.
- Excellent properties as a lubricant.
- It is the only alternative fuel that can be used in a conventional diesel engine, without modifications.
- Used cooking oils and fat residues from meat processing may be used as raw materials.

2.1.2 Disadvantages of the Use of Biodiesel

There are certain disadvantages of using biodiesel as a replacement for diesel fuel that must be taken into consideration:

- Slightly higher fuel consumption due to the lower calorific value of biodiesel.
- Slightly higher nitrous oxide (NO_x) emissions than diesel fuel.
- Higher freezing point than diesel fuel. This may be inconvenient in cold climates.
- It is less stable than diesel fuel, and therefore long-term storage (more than six months) of biodiesel is not recommended.
- May degrade plastic and natural rubber gaskets and hoses when used in pure form, in which case replacement with Teflon® components is recommended.
- It dissolves the deposits of sediments and other contaminants from diesel fuel in storage tanks and fuel lines, which then are flushed away by the biofuel into the engine, where they can cause problems in the valves and injection systems. In consequence, the cleaning of tanks prior to filling with biodiesel is recommended.

It must be noted that these disadvantages are significantly reduced when biodiesel is used in blends with diesel fuel.

2.2 Raw Materials for Biodiesel Production

The raw materials for biodiesel production are vegetable oils, animal fats and short chain alcohols. The oils most used for worldwide biodiesel production are rapeseed (mainly in the European Union countries), soybean (Argentina and the United States of America), palm (Asian and Central American countries) and sunflower, although other oils are also used, including peanut, linseed, safflower, used vegetable oils, and also animal fats. Methanol is the most frequently used alcohol although ethanol can also be used.

Since cost is the main concern in biodiesel production and trading (mainly due to oil prices), the use of non-edible vegetable oils has been studied for several years with good results.

Besides its lower cost, another undeniable advantage of non-edible oils for biodiesel production lies in the fact that no foodstuffs are spent to produce fuel [4]. These and other reasons have led to medium- and large-scale biodiesel production trials in several countries, using non-edible oils such as castor oil, tung, cotton, jojoba and jatropha. Animal fats are also an interesting option, especially in countries with plenty of livestock resources, although it is necessary to carry out preliminary treatment since they are solid; furthermore, highly acidic grease from cattle, pork, poultry, and fish can be used.

Microalgae appear to be a very important alternative for future biodiesel production due to their very high oil yield; however, it must be taken into account that only some species are useful for biofuel production.

Although the properties of oils and fats used as raw materials may differ, the properties of biodiesel must be the same, complying with the requirements set by international standards.

2.2.1 Typical Oil Crops Useful for Biodiesel Production

The main characteristics of typical oil crops that have been found useful for biodiesel production are summarized in the following paragraphs [6–10].

2.2.1.1 Rapeseed and Canola

Rapeseed adapts well to low fertility soils, but with high sulfur content. With a high oil yield (40–50%), it may be grown as a winter-cover crop, allows double cultivation and crop rotation.

It is the most important raw material for biodiesel production in the European Community. However, there were technological limitations for sowing and harvesting in some Central and South American countries, mainly due to the lack of adequate information about fertilization, seed handling, and storage (the seeds are very small and require specialized agricultural machinery). Moreover, low prices in comparison to wheat (its main competitor for crop rotation) and low production per unit area have limited its use.

Rapeseed flour has high nutritional value, in comparison to soybean; it is used as a protein supplement in cattle rations.

Sometimes canola and rapeseed are considered to be synonymous; canola (Canadian oil low acid) is the result of the genetic modification of rapeseed in the past 40 years, in Canada, to reduce the content of erucic acid and glucosinolates in rapeseed oil, which causes inconvenience when used in animal and human consumption.

Canola oil is highly appreciated due to its high quality, and with olive oil, it is considered as one of the best for cooking as it helps to reduce blood cholesterol levels.

2.2.1.2 Soybean

It is a legume originating in East Asia. Depending on environmental conditions and genetic varieties, the plants show wide variations in height. Leading soybean producing countries are the United States, Brazil, Argentina, China, and India.

Biodiesel production form soybean yields other valuable sub-products in addition to glycerin: soybean meal and pellets (used as food for livestock) and flour (which have a high content of lecithin, a protein). Grain yield varies between 2,000 and 4,000 kg/hectare. Since the seeds are very rich in protein, oil content is around 18%.

2.2.1.3 Oil Palm

Oil palm [11] is a tropical plant that reaches a height of 20–25 m with a life cycle of about 25 years. Full production is reached 8 years after planting.

Two kinds of oil are obtained from the fruit: palm oil proper, from the pulp, and palm kernel oil, from the nut of the fruit (after oil extraction, palm kernel cake is used as livestock food). Several high oil-yield varieties have been developed. Indonesia and Malaysia are the leading producers.

International demand for palm oil has increased steadily during the past years, the oil being used for cooking, and as a raw material for margarine production and as an additive for butter and bakery products.

It is important to remark that pure palm oil is semisolid at room temperature (20–22°C), and in many applications is mixed with other vegetable oils, sometimes partially hydrogenated.

2.2.1.4 Sunflower

Sunflower "seeds" are really a fruit, the inedible wall (husk) surrounding the seed that is in the kernel.

The great importance of sunflower lies in the excellent quality of the edible oil extracted from its seeds. It is highly regarded from the point of view of nutritional quality, taste and flavor. Moreover, after oil extraction, the remaining cake is used as a livestock feed. It must be noted that sunflower oil has a very low content of linoleic acid, and therefore it may be stored for long periods.

Sunflower adapts well to adverse environmental conditions and does not require specialized agricultural equipment and can be used for crop rotation with soybean and corn. Oil yield of current hybrids is in the range 48–52%.

2.2.1.5 Peanut

The quality of peanut is strongly affected by weather conditions during the harvest. Peanuts are mainly used for human consumption, in the manufacture of peanut butter, and as an ingredient for confectionery and other processed foods. Peanuts of lower quality (including the rejects from the confectionery industry) are used for oil production, which has a steady demand in the international market. Peanut oil is used in blends for cooking and as a flavoring agent in the confectionery industry.

The flour left over, following oil extraction, is of high quality with high protein content; in pellet form, it is used as a livestock feed.

2.2.1.6 Flax

Flax [12] is a plant of temperate climates, with blue flowers. Linen is made with the threads from the stem of the plant and the oil from the seeds is called linseed oil, used in paint manufacture. Flax seeds have nutritional value for human consumption since they are a source of polyunsaturated fatty acids necessary for human health. Moreover, the cake left over, following oil extraction, is used as a livestock feed.

The plant adapts well to a wide range of temperature and humidity; however, high temperatures and plentiful rain do not favor high yields of seed and fiber.

Flax seeds contain between 30 and 48% of oil, and protein content is between 20 and 30%. It is important to remark that linseed oil is rich in polyunsaturated fatty acids, linolenic acid being from 40 to 68% of the total.

2.2.1.7 Safflower

Safflower adapts well to dry environments. Although the grain yield per hectare is low, the oil content of the seed is high, from 30 to 40%. Therefore, it has economic potential for arid regions. Currently, safflower is used in oil and flour production and as bird feed.

There are two varieties, one rich in mono-unsaturated fatty acids (oleic acid) and the other with a high percentage of polyunsaturated fatty acids (linoleic acid). Both varieties have a low content of saturated fatty acids.

The oil from safflower is of high quality and low in cholesterol content. Other than being used for human consumption, it is used in the manufacture of paints and other coating compounds, lacquers and soaps.

It is important to note that safflower oil is extracted by means of hydraulic presses, without the use of solvents, and refined by conventional methods, without anti-oxidant additives.

The flour from safflower is rich in fiber and contains about 24% proteins. It is used as a protein supplement for livestock feed.

2.2.1.8 Castor Seed

The castor oil plant grows in tropical climates, with temperatures in the range 20–30°C; it cannot endure frost. It is important to note that once the seeds start germinating, the temperature must not fall below 12°C. The plant needs a warm and humid period in its vegetative phase and a dry season for ripening and harvesting. It requires plenty of sunlight and adapts well to several varieties of soils. The total rainfall during the growth cycle must be in the range 700–1,400 mm; although it is resistant to drought, the castor oil plant needs at least 5 months of rain during the year.

Castor oil is a triglyceride, ricinolenic acid being the main constituent (about 90%). The oil is non-edible and toxic owing to the presence of 1–5% of ricin, a toxic protein that can be removed by cold pressing and filtering. The presence of hydroxyl groups in its molecules makes it unusually polar as compared to other vegetable oils.

2.2.1.9 Tung

Tung [12] is a tree that adapts well to tropical and sub-tropical climates. The optimum temperature for tung is between 18 and 26°C, with low yearly rainfall.

During the harvest season, the dry nuts fall off from the tung tree and are collected from the ground. Nut production starts 3 years after the planting.

The oil from tung nuts is non-edible and used in the manufacture of paints and varnishes, especially for marine use.

2.2.1.10 Cotton

Among non-foodstuffs, cotton is the most widely traded commodity. It is produced in more than 80 countries and distributed worldwide. After the harvest, it may be traded as raw cotton, fiber or seeds. In cotton mills, fiber and seeds are separated from raw cotton.

Cotton fiber is processed to produce fabric and thread, for use in the textile industry. In addition, cotton oil and flour are obtained from the seed; the latter is rich in protein and is used in livestock feed and after further processing, for human consumption.

2.2.1.11 Jojoba

Although jojoba can survive extreme drought, it requires irrigation to achieve an economically viable yield.

Jojoba needs a warm climate, but a cold spell is necessary for the flowers to mature. Rainfall must be very low during the harvest season (summer). The plant reaches its full productivity 10 years after planting.

The oil from jojoba is mainly used in the cosmetics industry; therefore, its market is quickly saturated.

2.2.1.12 Jatropha

Jatropha is a shrub that adapts well to arid environments. *Jatropha curcas* is the most known variety; it requires little water or additional care; therefore, it is adequate for warm regions with little fertility. Productivity may be reduced by irregular rainfall or strong winds during the flowering season. Yield depends on climate, soil, rainfall and treatment during sowing and harvesting. Jatropha plants become productive after 3 or 4 years, and their lifespan is about 50 years.

Oil yield depends on the method of extraction; it is 28–32% using presses and up to 52% by solvent extraction. Since the seeds are toxic, jatropha oil is non-edible. The toxicity is due to the presence of curcasin (a globulin) and jatrophic acid (as toxic as ricin).

2.2.1.13 Avocado

Avocado is a tree between 5 and 15 m in height. The weight of the fruit is between 120 and 2.5 kg and the harvesting period varies from 5 to 15 months. The avocado fruit matures after picking and not on the tree.

Oil may be obtained from the fruit pulp and pit. It has a high nutritional value, since it contains essential fatty acids, minerals, protein and vitamins A, B6, C, D, and E. The content of saturated fatty acids in the pulp of the fruit and in the oil is low; on the contrary, it is very high in mono-unsaturated fatty acids (about 96% being oleic acid). The oil content of the fruit is in the range 12–30%.

2.2.1.14 Microalgae

Microalgae have great potential for biodiesel production, since the oil yield (in liters per hectare) could be one to two orders of magnitude higher than that of other raw materials. Oil content is usually from 20 to 50%, although in some species it can be higher than 70% [13]. However, it is important to note that not all microalgae are adequate for biodiesel production.

High levels of CO_2, water, light, nutrients and mineral salts are necessary for the growth of microalgae. Production processes take place in raceway ponds and photobiological reactors [13].

Leading oil crops used in biodiesel production are indicated in Box 2.1.

Box 2.1 Leading Oil Crops for Biodiesel Production

Rapeseed
Palm
Soybean

Fig. 2.1 Approximate oil yields for different crops

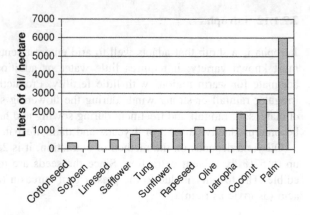

Figure 2.1 presents approximate oil-yield values (in liters per hectare) for some of the crops [13] discussed in this chapter.

It is important to note that the data in Fig. 2.1 only show the oil yields of different crops. However, for the comparison of economical suitability it must be borne in mind that in addition to oil, some crops are grown for fiber or protein production. For instance, soybean has an oil content of 18% (maximum), whereas the remainder is mostly protein (usually used as livestock feed).

2.2.2 Characteristics of Oils and Fats Used in Biodiesel Production

Oils and fats, known as lipids, are hydrophobic substances insoluble in water and are of animal or vegetal origin. They differ in their physical states at room temperature. From a chemical viewpoint, lipids are fatty glycerol esters known as triglycerides. The general chemical formula is shown in Fig. 2.2.

In Fig. 2.2, R_1, R_2 y R_3 represent hydrocarbon chains of fatty acids, which in most cases vary in length from 12 to 18 carbon atoms. The three hydrocarbon chains may be of equal or different lengths, depending on the type of oil; they may also differ on the number of double-covalent bonds in each chain.

Fatty acids may be saturated fatty acids (SFA) or non-saturated fatty acids (NSFA). In the former, there are only single covalent bonds in the molecules. The

Fig. 2.2 General chemical formula of triglycerides

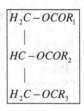

$$H_2C - OCOR_1$$
$$|$$
$$HC - OCOR_2$$
$$|$$
$$H_2C - OCR_3$$

Table 2.1 Chemical formulas of the main fatty acids in vegetable oils

Fatty acid	Chemical formula
Lauric (12:0)	$CH_3 (CH_2)_{10} COOH$
Palmitic (16:0)	$CH_3 (CH_2)_{14} COOH$
Estearic (18:0)	$CH_3 (CH_2)_{16} COOH$
Oleic (18:1)	$CH_3 (CH_2)_7 CH = CH (CH_2)_7 COOH$
Linoleic (18:2)	$CH_3 (CH_2)_4 CH = CH CH_2 CH = CH (CH_2)_7 COOH$
Linolenic (18:3)	$CH_3 CH_2 (CH = CH CH2)_3 (CH2)_6 COOH$
Erucic (22:1)	$CH_3 (CH_2)_7 CH = CH (CH2)_{11} COOH$
Ricinoleic (18:1)	$CH_3 (CH_2)_5 CHOH CH_2 CH = CH (CH_2)_7 COOH$

Table 2.2 Approximate content (in weight) of saturated and non-saturated fatty acids in some vegetable oils and animal fats

Oil/fat	SFA ($\approx$ % w/w)	NSFA ($\approx$ % w/w)
Coconut	90	10
Corn	13	87
Cottonseed	26	74
Olive	14	86
Palm	49	51
Peanut	17	83
Rapeseed	6	94
Soybean	14	86
Sunflower	11	89
Safflower	9	91
Castor	2	98
Yellow grease	33	67
Lard	41	59
Beef tallow	48	52

names of the most important fatty acids in oils are listed in Table 2.1 along with their chemical formulas [14]. The notation x:y indicates the number of carbon atoms in the oil molecule (x) and the number of unsaturations, i.e. double-covalent bonds (y). For instance, y = 0 for all the SFAs. Table 2.2 indicates the approximate content (in weight) of saturated and non-saturated fatty acids in some vegetable oils and animal fats.

The most frequent fatty acids in oils are lauric, palmitic, estearic, linoleic and linolenic, although others may also be present. It is important to note that vegetable oils differ in their content of fatty acids. For instance, ricinoleic acid is the main component of castor oil, whereas in olive oil it is oleic acid, in soybean oil it is linoleic acid, and in linseed oil it is linolenic acid.

The compositions indicated in Table 2.2 do not discriminate between the different saturated or unsaturated fatty acids. For instance, coconut oil has about 90% of SFAs in its composition (more than half being lauric acid), and palm oil has

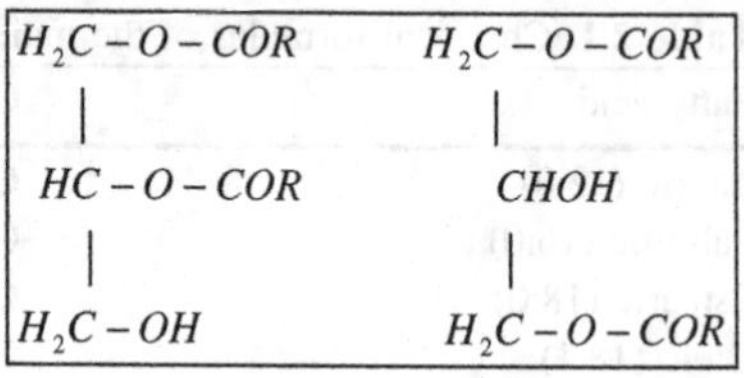

Fig. 2.3 Chemical formula
of diglycerides

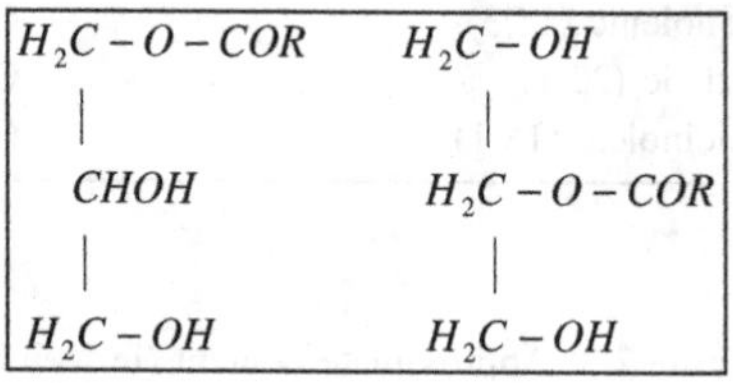

$$H_2C - O - COR \qquad\qquad H_2C - O - COR$$
$$|\qquad\qquad\qquad\qquad\qquad |$$
$$HC - O - COR \qquad\qquad CHOH$$
$$|\qquad\qquad\qquad\qquad\qquad |$$
$$H_2C - OH \qquad\qquad\qquad H_2C - O - COR$$

Fig. 2.4 Chemical formula
of monoglycerides

$$H_2C - O - COR \qquad\qquad H_2C - OH$$
$$|\qquad\qquad\qquad\qquad\qquad |$$
$$CHOH \qquad\qquad\qquad H_2C - O - COR$$
$$|\qquad\qquad\qquad\qquad\qquad |$$
$$H_2C - OH \qquad\qquad\qquad H_2C - OH$$

about 49% SFAs (more than 80% palmitic acid). Similarly, 60% of NSFAs content in soybean oil is linoleic acid, while in peanut more than 50% is oleic.

The US Department of Energy [15] indicates that a perfect biodiesel should only comprise mono-unsaturated fatty acids.

Vegetable oils may also contain small percentages of monoglycerides and diglycerides. Their chemical formulae are shown in Figs. 2.3 and 2.4. In addition, there will also be small amounts of free fatty acids (in most vegetable oils, less than 1%, except for palm oil, where they can reach up to 15%).

The composition of vegetable oils influences their properties [16]. For instance, the pour point and cloud point temperatures, cetane number and the iodine index depend on the number of unsaturations and the length of the fatty acid chains. A higher content of double-covalent bonds gives a lower solidification point and a higher iodine index.

2.2.3 Characteristics of Alcohols Used in Biodiesel Production

Alcohols that can be used in biodiesel production are those with short chains, including methanol, ethanol, butanol, and amylic alcohol. The most widely used alcohols are methanol (CH_3OH) and ethanol (C_2H_5OH) because of their low cost and properties. Methanol is often preferred to ethanol in spite of its high toxicity because its use in biodiesel production requires simpler technology; excess alcohol may be recovered at a low cost and higher reaction speeds are reached. The comparison between the two alcohols is summarized in Box 2.2.

It must be remembered that in order for biodiesel to be a fully renewable fuel, it should be obtained from vegetable oils and animal fats, together with an alcohol that is produced from biomass, such as bioethanol, instead of being a petro-chemical product. Several countries are carrying out research towards this objective, such as Spain and Brazil.

Box 2.2 Most Important Alcohols Used in Biodiesel Production

Methanol. Most widely used, in spite of its toxicity. It is a substance of petrochemical origin.
Ethanol. Less used, requires more complex production technology and the reaction speeds are lower. It can be produced from biomass.

2.3 Biodiesel Production Process

Biodiesel is produced from vegetable oils or animal fats and an alcohol, through a transesterification reaction [1, 2, 4, 5]. This chemical reaction converts an ester (vegetable oil or animal fat) into a mixture of esters of the fatty acids that makes up the oil (or fat). Biodiesel is obtained from the purification of the mixture of fatty acid methyl esters (FAME). A catalyst is used to accelerate the reaction (Fig. 2.5). According to the catalyst used, transesterification can be basic, acidic or enzymatic, the former being the most frequently used, as indicated in Box 2.3.

Box 2.3 Transesterification Reactions for Biodiesel Production

Basic. Most frequently used at all production scales.
Acid. Less frequent in industrial production, sometimes used a first stage with highly acidic raw materials.
Enzymatic. Less used; the enzymes are usually lipases.

A generic transesterification reaction is presented in Eq. (2.1); RCOOR$'$ indicates an ester, R$''$OH an alcohol, R$'$OH another alcohol (glycerol), RCOOR$''$ an ester mixture and *cat* a catalyst:

$$\text{RCOOR}' + \text{R}''\text{OH} \overset{\text{cat}}{\Leftrightarrow} \text{R}'\text{OH} + \text{RCOOR}''\qquad(2.1)$$

When methanol is the alcohol used in the transesterification process, the product of the reaction is a mixture of methyl esters; similarly, if ethanol were

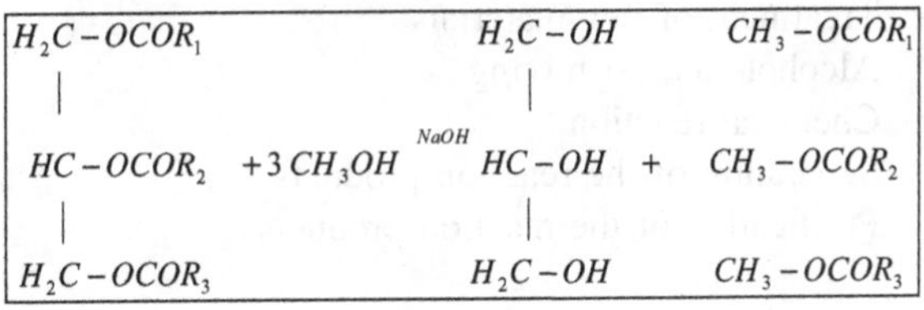

Fig. 2.5 Basic transesterification reaction with methanol

$$
\begin{array}{l}
H_2C-OCOR_1 \\
\quad | \\
HC-OCOR_2 \quad +3\,CH_3OH \overset{NaOH}{\longrightarrow} \\
\quad | \\
H_2C-OCOR_3
\end{array}
\quad
\begin{array}{l}
H_2C-OH \\
\quad | \\
HC-OH \quad + \\
\quad | \\
H_2C-OH
\end{array}
\quad
\begin{array}{l}
CH_3-OCOR_1 \\
\\
CH_3-OCOR_2 \\
\\
CH_3-OCOR_3
\end{array}
$$

$$
\begin{array}{lllll}
H_2C-OCOR_1 & & & H_2C-OH & CH_3CH_2-OCOR_1 \\
\ | & & & \ | & \\
HC-OCOR_2 & +3\,CH_3CH_2OH & \overset{NaOH}{\longrightarrow} & HC-OH & + & CH_3CH_2-OCOR_2 \\
\ | & & & \ | & \\
H_2C-OCOR_3 & & & H_2C-OH & CH_3CH_2-OCOR_3
\end{array}
$$

Fig. 2.6 Basic transesterification reaction with ethanol

used, the reaction product would be a mixture of ethyl esters. In both cases, glycerin will be the co-product of the reaction. This is shown schematically in Figs. 2.5 and 2.6.

Although transesterification is the most important step in biodiesel production (since it originates the mixture of esters), additional steps are necessary to obtain a product that complies with international standards [4, 17], as shown in Box 2.4. In consequence, once the chemical reaction is completed and the two phases (mix of esters and glycerin) are separated, the mix of methyl esters must be purified to reduce the concentration of contaminants to acceptable levels. These include remnants of catalyst, water and methanol; the latter is usually mixed in excess proportion with the raw materials in order to achieve higher conversion efficiency in the transesterification reaction. In the following sections the steps of the purification process will be described in detail.

2.3.1 Treatment of Raw Materials

The content of free fatty acids, water and non-saponificable substances are key parameters to achieve high conversion efficiency in the transesterification reaction.

The use of basic catalysts in triglycerides with high content of free fatty acids is not advisable [18], since part of the latter reacts with the catalyst to form soaps. In consequence, part of the catalyst is spent, and it is no longer available for transesterification. In summary the efficiency of the reaction diminishes with the increase of the acidity of the oil; basic transesterification is viable if the content of free fatty acids (FFAs) is less than 2%. In the case of highly acidic raw materials

Box 2.4 Stages of Biodiesel Production Process

Treatment of raw materials
Alcohol-catalyst mixing
Chemical reaction
Separation of the reaction products
Purification of the reaction products

(animal fats from cattle, poultry, pork; vegetable oils from cotton, coconut, most used oils, etc.) an acid transesterification [19] is necessary as a preliminary stage, to reduce the level of FFAs to the above-mentioned value.

Besides having low humidity and acid content, it is important that the oil presents a low level of non-saponificable substances. If the latter were to be present in significant amounts and soluble in biodiesel, it would reduce the level of esters in the product, making it difficult to comply with the minimum ester content required by the standards.

The AOCS standards [20] list the required properties of oils. Anyway, the properties required by the oils are finally determined by the biodiesel industry in each country. For instance, in Argentina the oils for biodiesel production usually have:

- Acidity level <0.1 mg KOH/g
- Humidity <500 ppm
- Peroxide index <10 meq/kg
- Non-saponificable substances <1%.

2.3.2 Alcohol-Catalyst Mixing

The alcohol used for biodiesel production must be mixed with the catalyst before adding the oil. The mixture is stirred until the catalyst is completely dissolved in the alcohol. It must be noted that the alcohol must be water-free (anhydrous) for the reasons explained in the previous paragraph.

Sodium and potassium hydroxides are among the most widely used basic catalysts. For production on an industrial scale, sodium or potassium methoxides or methylates are commercially available.

Of course, due caution must be exercised, and all applicable safety regulations must be followed, when working with methanol, hydroxides and methoxides, independently of the production scale.

The alcohol-to-oil volume ratio, R, is another key variable of the transesterification process. The stoichiometric ratio (Fig. 2.5) requires 1 mol of oil to react with 3 mol of alcohol, to obtain 3 mol of fatty acids methyl esters (FAME) and 1 mol of glycerin. However, since the reaction is reversible, excess alcohol as a reactant will shift the equilibrium to the right side of the equation, increasing the amount of products (as it may be inferred from Le Chatelier's principle). Although a high alcohol-to-oil ratio does not alter the properties of FAME, it will make the separation of biodiesel from glycerin more difficult, since it will increase the solubility of the former in the latter. Usually, a 100% alcohol excess is used in practice, that is, 6 mol of alcohol per mole of oil. This corresponds to a 1:4 alcohol-to-oil volume ratio ($R = 0.25$). The relation between the dielectric properties of FAME and the alcohol-to-oil ratio, R, will be discussed in Chap. 7.

Finally, it must be noted that the necessary amount of catalyst is determined taking into account the acidity of the oil, by titration.

2.3.3 Chemical Reaction

The chemical reaction takes place when the oil is mixed with the alkoxide (alcohol–catalyst mix) described in the previous paragraph. This requires certain conditions [17, 21] of time, temperature and stirring. Since alcohols and oils do not mix at room temperature, the chemical reaction is usually carried out at a higher temperature and under continuous stirring, to increase the mass transfer between the phases.

Usually, emulsions form during the course of the reaction; these are much easier and quicker to destabilize when methanol is used, in comparison to ethanol [22]. Due to the greater stability of emulsions formed, difficulties arise in the phase separation and purification of biodiesel when ethanol is used in the reaction.

The transesterification process may be carried out at different temperatures. For the same reaction time, the conversion is greater at higher temperatures. Since the boiling point of methanol is approximately 68°C (341 K), the temperature for transesterification at atmospheric pressure is usually in the range between 50 and 60°C.

It is very useful to know the chemical composition of the mixture during the reaction; then, if the reaction mechanism and kinetics are known, the process can be optimized. However, the determination of the mixture composition is not easy, since more than a hundred substances are known to be present [23]. For instance, for biodiesel production from rapeseed oil (whose main SFAs are palmitic, oleic, linoleic and linolenic) and methanol, with potassium hydroxide as a catalyst, it could be theoretically possible to find 64 isomers of triglycerides, 32 diglycerides, 8 monoglycerides, their methyl esters, potassium salts of the fatty acids, potassium methoxide, water, etc.

The studies on this subject [24, 25] indicate the following general guidelines:

- For longer reaction times, the concentration of triglycerides diminishes the concentration of esters, increases and the concentration of mono- and diglycerides increases to a maximum and then decreases.
- Most of the chemical reaction takes place during the first minutes.
- The absence of mono- and diglycerides at the beginning of the chemical reaction and the increase and reduction of their concentration during the reaction confirm that the production of esters from the triglycerides takes place in three steps, as represented in the equations below:

$$TG + MOH \xrightarrow{KOH} DG + ME \qquad (2.2)$$

$$DG + MOH \xrightarrow{KOH} MG + ME \qquad (2.3)$$

$$MG + MOH \xrightarrow{KOH} G + ME \qquad (2.4)$$

where MOH indicates methanol, ME are the methyl esters, TG, DG and MG are tri-, di- and monoglycerides, respectively, and G is the glycerin.

Several methods, with different levels of equipment complexity and training requirements, have been devised to analyze samples that are mixtures of fatty acids and esters from mono-, di-, and triglycerides obtained from transesterification of vegetable oils. A list is presented in Box. 2.5.

It must be noted that thin layer chromatography (TLC) provides essentially qualitative information about the sample composition, as distinct from the other methods in Box 2.5, that can be used for quantitative analysis. However, the simplicity, speed and low cost of TLC make it quite attractive as a technique for process optimization and routine checks, especially in small- and medium-scale production plants, and also for training purposes.

2.3.3.1 Catalysts

The catalysts used for the transesterification of triglycerides may be classified as basic, acid or enzymatic, as indicated in Box 2.3 [38, 39].

Basic catalysts include sodium hydroxide (NaOH), potassium hydroxide (KOH), carbonates and their corresponding alcoxides (for instance, sodium methoxide or ethoxide). There are many references on basic catalysts in the scientific literature [26, 40–51].

Acid catalysts include sulfuric acid, sulfonic acids and hydrochloric acid; their use has been less studied [26, 52–57].

Heterogeneous catalysts that have been considered for biodiesel production include enzymes [39], titanium silicates [58], and compounds from alkaline earth metals [59], anion exchange resins [59] and guanidines in organic polymers [60]. Lipases are the most frequently used enzymes for biodiesel production [61–64].

2.3.4 Separation of the Reaction Products

The separation of reaction products takes place by decantation: the mixture of fatty acids methyl esters (FAME) separates from glycerin forming two phases, since

---

Box 2.5 Analytical Methods for Mixtures of Fatty Acid and Esters

Thin Layer Chromatography [26]
Gas Chromatography (GC) [27–29]
High Performance Liquid Chromatography (HPLC) [29–33]
Gel Permeation Chromatography [34]
Nuclear Magnetic Resonance (NMR) [35, 36]
Infrared (IR) Spectroscopy [36, 37]

they have different densities; the two phases begin to form immediately after the stirring of the mixture is stopped. Due to their different chemical affinities, most of the catalyst and excess alcohol will concentrate in the lower phase (glycerin), while most of the mono-, di-, and triglycerides will concentrate in the upper phase (FAME). Once the interphase is clearly and completely defined, the two phases may be physically separated. It must be noted that if decantation takes place due to the action of gravity alone, it will take several hours to complete. This constitutes a "bottleneck" in the production process, and in consequence the exit stream from the transesterification reactor is split into several containers. Centrifugation is a faster, albeit more expensive alternative.

After the separation of glycerin, the FAME mixture contains impurities such as remnants of alcohol, catalyst and mono-, di-, and triglycerides. These impurities confer undesirable characteristics to FAME, for instance, increased cloud point and pour point, lower flash point, etc. In consequence a purification process is necessary for the final product to comply with standards. This will be discussed in the next section.

2.3.5 Purification of the Reaction Products

The mixture of fatty acids methyl esters (FAME) obtained from the transesterification reaction must be purified in order to comply with established quality standards for biodiesel. Therefore, FAME must be washed, neutralized and dried.

Successive washing steps with water remove the remains of methanol, catalyst and glycerin, since these contaminants are water-soluble. Care must be taken to avoid the formation of emulsions during the washing steps, since they would reduce the efficiency of the process. The first washing step is carried out with acidified water, to neutralize the mixture of esters. Then, two additional washing steps are made with water only. Finally the traces of water must be eliminated by a drying step. After drying, the purified product is ready for characterization as biodiesel according to international standards.

An alternative to the purification process described above is the use of ion exchange resins or silicates.

Glycerin as obtained from the chemical reaction is not of high quality and has no commercial value. Therefore, it must be purified after the phase separation. This is not economically viable in small scale production, due to the small glycerin yield. However, purification is a very interesting alternative for large-scale production plants, since, in addition to the high quality glycerin, part of the methanol is recovered for reutilization in the transesterification reaction (both from FAME and glycerin), and thus lowering biodiesel production costs. The steady increase of biodiesel production is fostering research for novel uses of glycerin in the production of high-value-added products.

It must be noted that the stages of the biodiesel production process (summarized in Box 2.4) are the same for all the production scales (laboratory, pilot plant, small-, medium-, and large-scale industrial). However, the necessary equipment will be significantly different [2].

2.4 Glycerin

Glycerin is the usual name of 1,2,3-propanetriol; it is also referred to as glycerol, glycerin or glycyl alcohol. Chemically an alcohol, it is a liquid of high viscosity at room temperature, odorless, transparent, colorless, of low toxicity and sweet taste. The boiling point of glycerin is high, 290°C (563 K), and its viscosity increases noticeably at low temperature, down to its freezing point, 18°C (291 K). It is a polar substance that can be mixed with water and alcohols, and is also a good solvent. Glycerin is hygroscopic and has humectant properties.

Until the last years of the nineteenth century, glycerin was a product of candle manufacturing (from animal fat) and it was used mainly in the production of nitroglycerin for explosives. Later, separation processes from soap were developed and most glycerin was obtained as a sub-product of the soap industry. Since mid-century, synthetic glycerin can also be obtained using raw materials from the petrochemical industry.

At present, glycerin is obtained as a sub-product of soap or biodiesel production, and it is purified to eliminate the contaminants, mainly partially dissolved soap or salt (for the sub-product of soap production), or catalyst and methanol (from biodiesel production).

2.4.1 Uses

Even though the main use of glycerin was traditionally in the soap industry, about the middle of the twentieth century more than 1,500 uses for glycerin had been identified. These include the manufacture, conservation, softening and moisturizing of an ample variety of products [65]. Some of the uses of glycerin are:

- As an additive in the manufacture of soaps, to improve their properties
- In the manufacture of nitroglycerin for the production of explosives
- In the food industry, for the manufacture of sweets, soft drinks, and pet foods and in the conservation of canned fruit
- Due to its moisturizing and emollient properties, in the cosmetics industry for the manufacture of creams and lotions
- In the chemical industry, for the fabrication of urethane foams, alkydic resins and cellophane, among other uses
- In the pharmaceutical industry, for the manufacture of ointments, creams and lotions
- In the manufacture of certain inks
- For the lubrication of molds.

In the last years, glycerin production has increased, due to the steady growth of biodiesel production. Several academic and industrial research groups are actively pursuing new applications for glycerin, particularly in connection with polymers and surfactants [65]. It must be noted that the uses of glycerin are in principle similar to those of other widely used poliols (glycol, sorbitol, pentaerythritol, etc.), thus opening the technological possibility of replacing these poliols in new applications. Of course, these substitutions will take place if economically viable, and will depend on the prices of glycerin and the poliols involved.

The main research objective is the production of high-value-added products using glycerin, for instance, as a substrate for protein production from single-cell organisms, as a raw material for the production of detergents and bioemulsifiers, for the production of other poliols by fermentation (such as 1,2-propanediol or 1,3-propanediol), or for the production of other biofuels (bioethanol, biogas, hydrogen).

2.5 Concluding Remarks

There are significant advantages in the use of biodiesel as a replacement of diesel fuel and in blends.

The vegetable oils used as raw materials can be obtained from different oil crops that may be grown in a wide variety of environments, some of which are not adequate for traditional agricultural production. Microalgae grown in ponds and photobiological reactors have also great potential for the production of oils for biodiesel production. Moreover, used cooking oils and fat residues from the meat processing industry may also be employed in biodiesel production.

The production process has the same stages, irrespective of the production scale, although the differences in equipment may be significant. After the treatment of the raw materials, the transesterification reaction (usually with methanol and a basic catalyst) produces a mixture of fatty acids methyl esters (FAME) with glycerin as a co-product. The mixture of methyl esters must be separated from the glycerin and purified in order to comply with the requirements set by international standards for biodiesel.

In large-scale production plants, glycerin is usually recovered and purified since it is a valuable substance, with many applications in the pharmaceutical, cosmetics and chemical industries.

References

1. Knothe G, Dunn RO, Bagby MO (1997) Biodiesel: the use of vegetable oils and their derivatives as alternative diesel fuels. In: Fuels and Chemicals from Biomass, 1st edn. American Chemical Society, New York
2. Van Gerpen J, Shanks B, Pruszko R, Clements D, Knothe G (2004) Biodiesel production technology. National Renewable Energy Laboratory, NRRL/SR-510-36244

3. Van Gerpen J, Shanks B, Pruszko R, Clements D, Knothe G (2004) Biodiesel analytical methods. National Renewable Energy Laboratory, NRRL/SR-510-36240
4. Romano SD, González Suárez E, Laborde MA (2006) Biodiesel. In: Combustibles Alternativos, 2nd edn. Ediciones Cooperativas, Buenos Aires
5. Fukuda H, Kondo A, Noda H (2001) Biodiesel fuel production by transesterification of oils. J Biosci Bioeng 92(5):405–416
6. Romano SD, González Suárez E (2009) Biocombustibles líquidos en Iberoamérica, 1st edn. Ediciones Cooperativas, Buenos Aires
7. Secretaría de Agricultura, Ganadería, Pesca y Alimentos, República Argentina http://www.sagpya.mecon.gov.ar/new/0-0/agricultura/otros/estimaciones/infgeneral.php. Accessed 6 November 2009
8. Secretaría de Agricultura, Ganadería, Pesca y Alimentos, República Argentina http://www.biodiesel.gov.ar. Accessed 6 November 2009
9. Empresa Brasileira de Pesquisa Agropecuária http://www.embrapa.br/. Accessed 6 November 2009
10. Ministerio da Agricultura Pecuária e Abastecimento do Brazil http://www.agricultura.gov.br/. Accessed 6 November 2009
11. Agudelo Santamaría J R, Benjumea Hernández P N (2004) Biodiesel de aceite crudo de palma colombiano: aspectos de su obtención y utilización
12. García Penela JM (2007) Selección de indicadores que permitan determinar cultivos óptimos para la producción de biodiesel en las ecoregiones Chaco Pampeanas de la República Argentina. INTA, Buenos Aires
13. Chisti Y (2007) Biodiesel from microalgae. Biotech Adv 25:294–306
14. Weast RC, Astle MJ, Beyer WH (1986) Handbook of Chemistry and Physics, 66 th edn. CRC Press, Florida
15. USA Department of Energy (2004) Biodiesel: handling and uses guidelines. Energy Efficiency and Renewable Energy
16. Pryde EH (1981) Vegetable oil versus diesel fuel: chemistry and availability of vegetable oils. In: Proceedings of regional workshops on alcohol and vegetable oil as alternative fuels
17. Meher LC, Vidya Sagar D, Naik SN (2006) Technical aspects of Biodiesel production by transesterificatiom: a review. Renew Sustain Energy Rev 10(3):248–268
18. Turck R (2002) Method for producing fatty acid ester of monovalent alkyl alcohols and use thereof. USP 0156305
19. Tomasevic AV, Marinkovic SS (2003) Methanolysis of used frying oils. Fuel Process Technol 81:1–6
20. Berner D (1989) AOCS' 4th edition of methods. J Am Oil Chem Soc 66(12):1749
21. Freedman B, Pryde EH, Mounts TL (1984) Variables affecting the yields of fatty esters from transesterified vegetable oils. J Am Oil Chem Soc 61(19):1638–1643
22. Zhou W, Konar SK, Boocock DGV (2003) Ethyl esters from the single-phase base-catalysed ethanolysis of vegetable oils. J Am Oil Chem Soc 80(4):367–371
23. Komers K, Stloukal R, Machek J, Skopal F (2001) Biodiesel from rapeseed oil, methanol and KOH. 3. Analysis of composition of actual mixture. Eu J Lipid Sci Technol 103(6):363–71
24. Komers K, Stloukal R, Machek J, Skopal F, Komersová A (1998) Biodiesel from rapeseed oil, methanol and KOH. Analytical methods in research and production. Fett/Lipid 100(11):507–512
25. Peter SKF, Ganswindt R, Neuner HP, Weidner E (2002) Alcoholysis of triacylglycerols by heterogeneous catalysis. Eur J Lipid Sci Technol 104(6):324–330
26. Freedman B, Pryde EH, Kwolek WH (1984) Thin layer chromatography/flame ionization analysis of transesterified vegetable oils. J Am Oil Chem Soc 61(7):1215–1220
27. Mittelbach M (1993) Diesel fuel from vegetable oils, V: gas chromatographic determination of free glycerol in transesterified vegetable oils. Chromatogr 37(11–12):623–626
28. Plank C, Lorbeer E (1995) Simultaneous determination of glycerol, mono-, di-, and triglycerides in vegetable oil methyl esters by capillary gas chromatography. J Chromatogr A 697:461–468

29. Knothe G (2001) Analytical methods used in the production and fuel quality assessment of biodiesel. Trans ASAE 44(2):193–200
30. Trathnigg B, Mittelbach M (1990) Analysis of triglyceride methanolysis mixtures using isocratic HPLC with density detection. J Liq Chromatogr 13(1):95–105
31. Lozano P, Chirat N, Graille J, Pioch D (1996) Measurement of free glycerol in biofuels. Fresenius J Anal Chem 354:319–322
32. Noureddini H, Zhu D (1997) Kinetics of transesterification of soybean oil. J Am Oil Chem Soc 74(11):1457–1463
33. Holcapek M, Jandera P, Fischer J, Prokes B (1999) Analytical monitoring of the production of biofuel by high performance liquid chromatography with various detection methods. J Chromatogr A 858:13–31
34. Darnoko D, Cheryan M, Perkins EG (2000) Analysis of vegetable oil transesterification products by gel permeation chromatography. J Liq Chrom Rel Technol 23(15):2327–2335
35. Gelbard G, Bres O, Vargas RM, Vielfaure F, Schuchardt UF (1995) 1H nuclear magnetic resonance determination of the yield of the transesterification of rapeseed oil with methanol. J Am Oil Chem Soc 72(10):1239–1241
36. Knothe G (2000) Monitoring a processing transesterification reaction by fibre-optic color near infrared spectroscopy with correlation to ^{1}H nuclear magnetic resonance. J Am Oil Chem Soc 77(5):489–493
37. Knothe G (1999) Rapid monitoring of transesterification and accessing biodiesel fuel quality by near-infrared spectroscopy using a fibre optic probe. J Am Oil Chem Soc 76(7):795–800
38. Ma F, Hanna MA (1999) Biodiesel production: a review. Biores Tech 70(1):1–15
39. Vicente G, Martínez M, Aracil J (2004) Integrated biodiesel production: a comparison of homogeneous catalyst systems. Bioresour Technol 92:297–305
40. Wright HJ, Segur JB, Clark HV, Coburn SK, Langdom EE, DuPuis RN (1944) Oil and Soap 21:145–148
41. Bradshaw GB, Meuly WC (1944) US Patent 2 360 844
42. Feuge RO, Gros AT (1949) Modification of vegetable oils. J Am Oil Chem Soc 26(3):97–102
43. Gauglitz EJ, Lehman LW (1963) The preparation of alkyl esters from highly unsaturated triglicerides. J Am Oil Chem Soc 40:197–198
44. Mittelbach M, Trathnigg B (1990) Kinetics of alkaline catalyzed methanolysis of sunflower oil. Fat Sci Technol 92(4):145–148
45. Nye MJ, Williamson TW, Deshpande S, Schrader JH (1983) Conversion of used frying oil to diesel fuel by transesterification: preliminary tests. J Am Oil Chem Soc 60(8):1598–1601
46. Schwab AW, Bagby MO, Freedman B (1987) Preparation and properties of diesel fuels from vegetable oils. Fuel 66(10):1372–1378
47. Peterson CL, Feldman M, Korus R, Auld DL (1991) Batch type transesterification process for winter rape oil. Appl Eng Agric 7(6):711–716
48. Boocock DGB, Konar SK, Mao V, Sidi H (1996) Fast one phase oil rich processes for the preparation of vegetable oil methyl esters. Biomass Bioenergy 11(1):43–50
49. Cvengros J, Povazanec F (1996) Production and treatment of rapeseed oil methyl esters as alternative fuels for diesel engines. Bioresour Technol 55:145–152
50. Noureddini H, Harkey D, Medikonduru V (1998) A countinuous process for the conversion of vegetable oils into methylesters of fatty acids. J Am Oil Chem Soc 75(12):1775–1783
51. Vicente G (2001) Study of the biodiesel production. PhD thesis, Facultad de Química. Universidad Complutense de Madrid
52. Conde Cotes A, Wenzel L (1974) Cinética de la transesterificación del aceite de higuerilla. Revista Latinoamericana de Ingeniería Química y Química Aplicada 4:125–141
53. Harrington KJ, D'Arcy-Evans C (1985) A comparison of conventional and in situ methods of transesterification of seed oil from a series of sunflower cultivars. J Am Oil Chem Soc 62(6):1009–1013
54. Özgül S, Türkay S (1993) In situ esterification of rice bran oil with methanol and ethanol. J Am Oil Chem Soc 70(2):145–147

55. Kildiran G, Yücel SÖ, Türhay S (1996) In situ-alcoholysis of soybean oil. J Am Oil Chem Soc 73(2):225–228
56. Marinkovic S, Tomasevic A (1998) Transesterification of sunflower oil in situ. Fuel 77(12):1389–1391
57. Canakci M, Van Germen J (1999) Biodiesel production via acid catalysis. Trans ASAE 1999, 42(5):1203–1210
58. Bayense CR (1994) Esterification process. European patent number 06 23581 A2
59. Peterson GR, Scarrah WP (1984) Rapeseed oil transesterification by heterogeneous catalysis. J Am Oil Chem Soc 61(10):1593–1597
60. Schuchardt U, Vargas RM, Gelbard G (1996) Transesterification of soybean oil catalyzed by alkylguanidines heterogenized on different substituted polystyrenes. J Mol Catal A Chem 109(1):37–44
61. Nelson LA, Foglia TA, Marmer WN (1996) Lipase-catalyzed production of biodiesel. J Am Oil Chem Soc 73(8):1191–1195
62. Mittelbach M (1990) Lipase catalyzed alcoholysis of sunflower oil. J Am Oil Chem Soc 67(3):168–170
63. Shimada Y, Watanabe Y, Samukawa T, Sugihara A, Noda H, Fukuda H, Tominaga Y (1999) Conversion of vegetable oil to biodiesel using immobilized Candida antarctica lipase. J Am Oil Chem Soc 76(7):789–793
64. Kaieda M, Samukawa T, Matsumoto T, Ban K, Kondo A, Shiada Y, Noda H, Nomoto F, Ohtsuka K, Izumoto E, Fukuda H (1999) Biodiesel fuel production from plant oil catalyzed by Rhizopus oryzae lipase in a water-containing system without an organic solvent. J Biosci Bioeng 88:627–631
65. Claude S (1999) Research of new outlets for glycerol—recent developments in France. Fett/Lippid 3:101–104

Chapter 3
Standards for Fuel Characterization

3.1 Introduction

Quality control during production and distribution is critical to ensure that the fuel reaching end users has a reliable and consistent quality and guarantees a good engine performance. To this end, the fuel is characterized by measuring different properties according to international standards.

3.2 Biodiesel Characterization According to International Standards

The properties of biodiesel depend on the proportion and type of methyl/ethyl esters that conform it; in other words, on the composition of fatty acids of the oil used as feedstock. This fluctuating proportion is the main reason by which norms provide ranges in some properties such as density or viscosity.

Austria was the first country that determined and adopted a quality standard for rapeseed oil methyl esters, in 1990 (ON C 1190).

Depending on which norm is taken as reference (EN, ASTM, DIN, ON, CSN, UNI, etc.) the standardized value ranges may vary, but in most of them the variations are minor.

Standards indicate the different properties that must be measured, the ranges where these measurements should lie to comply with specifications, the unit of measurement and the type and the specific norms under which specific properties must be measured. In some cases the standards indicate maximum limits (i.e., in the case of impurities) and/or minimum limits (e.g., for the degree of conversion of the reaction).

S. D. Romano and P. A. Sorichetti, *Dielectric Spectroscopy in Biodiesel Production and Characterization*, Green Energy and Technology, DOI: 10.1007/978-1-84996-519-4_3, © Springer-Verlag London Limited 2011

The parameters that define biodiesel quality can be classified into two major groups [1]: the first includes general properties (density, viscosity, flash point, cloud point, pour point, cetane number, neutralization number, etc.) and the second describes the chemical composition and purity of the mixture of fatty acid esters (alcohol content, ester content, mono-, di- and tri-glyceride content, total and free glycerol quantity, iodine value, etc.).

Within the European Community, the standard adopted is the Pr EN 14214/09 [2]. The characterization properties required by this standard are shown in Box 3.1 (specific test methods are listed in brackets).

Box 3.1 Properties Required for the Characterization of Biodiesel According to European Standard EN 14214/09

FAME content (EN 14103) [3]
Density at 15°C (EN ISO 3675, EN ISO 12185) [4, 5]
Viscosity at 40°C (EN ISO 3104) [6]
Flash point (EN ISO 2719, EN ISO 3679) [7, 8]
Sulfur content (EN ISO 20846, EN ISO 20884, WI 019 376) [9–11]
Cetane number (EN ISO 5165) [12]
Sulfated ash content (ISO 3987) [13]
Water content (EN ISO 12937) [14]
Total contamination (EN 12662) [15]
Copper strip corrosion at 3 h and 50°C (EN ISO 2160) [16]
Oxidation stability at 110°C (EN 15751, EN 14112, WI 019 370) [17–19]
Acid value (EN 14104) [20]
Iodine value (EN 14111) [21]
Linolenic acid methyl ester (EN 14103) [3]
Polyunsaturated (with ≥ 4 double bounds) methyl esters (EN 15779) [22]
Methanol content (EN 14110) [23]
Monoglyceride content (EN 14105) [24]
Diglyceride content (EN 14105) [24]
Triglyceride content (EN 14105) [24]
Free glycerol (EN 14105, EN 14106) [24, 25]
Total glycerol (EN 14105) [24]
Group I metals (Na and K) (EN 14108, EN 14109, EN 14538) [26–28]
Group II metals (Ca and Mg) (EN 14538) [28]
Phosphorus content (WI 019 376, EN 14107) [11, 29]
Cloud filter plugging point (EN 116) [30]
Note. The acronym FAME means fatty acid methyl esters.

On the other hand, US standard ASTM D 6751/09 [31] lists the properties indicated in Box 3.2.

Box 3.2 Properties Required for the Characterization of Biodiesel According to US Standard ASTM D 6751/09

Calcium and magnesium, combined (EN 14538) [28]
Flash point (closed cup) (ASTM D 93) [32]
Methanol content (EN 14110) [23]
Water and sediment (ASTM D 2709) [33]
Kinematic viscosity at 40°C (ASTM D 445) [34]
Sulfated ash (ASTM D 874) [35]
Sulfur (ASTM D 5453) [36]
Copper strip corrosion (ASTM D 130) [37]
Cetane number (ASTM D 613) [38]
Cloud point (ASTM D 2500) [39]
Carbon residue (ASTM D 4530) [40]
Acid number (ASTM D 664) [41]
Free glycerin (ASTM D 6584) [42]
Total glycerin (ASTM D 6584) [42]
Phosphorus content (ASTM D 4951) [43]
Distillation temperature, Atmospheric equivalent temperature (90% recovered) (ASTM D1160) [44]
Sodium and potassium, combined (EN 14538) [28]
Oxidation stability (EN 14112) [18]

Table 3.1 shows a comparison between ASTM D 6751/09 [31], Pr EN 14214/09 [2] and the previous European standard EN 14214/03 [45].

From the table it may be seen that Pr EN 14214/09 requires the determination of more properties than ASTM D 6751/09. Regarding the specific values of the parameters, main differences lie in phosphorus content, acid value, oxidation stability, cetane number and sulfur content. In all cases, values of ASTM D 6751/09 are less stringent than those indicated in Pr EN 14214/09.

It is interesting to point it out that the standard Pr EN 14214/09 is stricter than EN 14214/03 in the oxidation stability and phosphorus content properties. On the other hand, carbon residue determination has been eliminated, polyunsaturated (with ≥4 double bounds) methyl esters have been added and the value of minimum flash point has been reduced from 120 to 101°C. Moreover, several methods for determination of properties in Pr EN 14214/09 have been updated, and also the reduction of the maximum value of sodium and potassium from 5 to 3.5 mg/kg is under evaluation. The text in Pr EN 14214/09 indicates that "the specification has been improved in the perspective of blending at up to 10% (V/V) of FAME in regular fossil-based diesel".

ASTM D 6751/09 has incorporated the measurement of additional properties, such as calcium and magnesium, combined; sodium and potassium, combined; oxidation stability and methanol content. Test methods for determining calcium

Table 3.1 Comparison between properties listed on ASTM D 6751/09, EN 14214/03 and Pr EN 14214/09

Property	Unit	ASTM D/09 6751/09	EN/03 14214/03	Pr EN/09 14214/09
Ester content	% w/w	–	≥96.5	≥96.5
Density	kg/m^3	–	860–900	860–900
Kinematic viscosity	mm^2/s	1.9–6.0	3.5–5.0	3.5–5.0
Flash point	°C	≥130 ≥93 (closed cup)	≥120	≥101
Sulfur content	mg/kg	≤15	≤10	≤10
Carbon residue	% w/w	≤0.05	≤0.30	–
Cetane number		≥47	≥51	≥51
Sulfated ash content	% w/w	≤0.02	≤0.02	≤0.02
Water and sediment	% w/w	≤0.05	–	–
Water content	mg/kg	–	≤500	≤500
Total contamination	mg/kg	–	≤24	≤24
Copper strip corrosion		≤No. 3	Class 1	Class 1
Oxidation stability	h	≥3	≥6	≥8
Acid number	mg KOH/g	≤0.80	≤0.50	≤0.50
Iodine value	g iodine/100 g	–	≤120	≤120
Linolenic acid methyl ester	% w/w	–	≤12.0	≤12.0
Polyunsaturated methyl esters	% w/w	–	≤1	≤1
Methanol content	% w/w	≤0.20	≤0.20	≤0.20
Monoglyceride content	% w/w	–	≤0.80	≤0.80
Diglyceride content	% w/w	–	≤0.20	≤0.20
Triglyceride content	% w/w	–	≤0.20	≤0.20
Free glycerol	% w/w	≤0.020	≤0.020	≤0.020
Total glycerol	% w/w	≤0.24	≤0.25	≤0.25
Group I metals (sodium and potassium)	mg/kg	≤5.0	≤5.0	≤5.0
Group II metals (calcium and magnesium)	mg/kg	≤5.0	≤5.0	≤5.0
Phosphorus content	mg/kg	≤10.0	≤10.0	≤2.0
Cold soak filterability	s	≤360	–	–
Cold filter plugging point (CFPP)	°C	–	Depending on the grade	Depending on the grade

and magnesium, sodium and potassium, methanol content and oxidation stability are based on specific EN standards. According to ASTM D 6751/03, although the minimum flash point indicated in the Table 1 of the Standard ("Detailed Requirements for Biodiesel") is 130°C, the text (paragraph X1.2.3) states that "the flash point specification for biodiesel is intended to be 100°C minimum. Due to high variability with Test Method D 93 as the flash point approaches 100°C, the flash point specification has been set at 130°C minimum to ensure an actual value of 100°C minimum". On the other hand, the maximum value for cold filter plugging point (CFPP) depends on the grade. There are six grades for temperature climates and five different classes for arctic climates. The grades depend on the seasons and regions. The range for the maximum CFPP varies from 5°C (grade A) to −20°C (grade F).

When oil companies buy biodiesel to blend with diesel fuel, frequently they require stricter specifications. For instance, many oil companies in the European Union request the following values:

- Water content <300 ppm
- Total contamination <15 ppm
- Monoglyceride content <0.4% w/w
- Oxidation stability >8 h
- FBT (filter blocking tendency by IP387) <2
- Sterolglucosides <20 ppm.

3.3 Characterization Properties Required by Standards

In this section, the main properties required by Standards will be defined [31]. Also, a brief explanation will be given of the inconveniences for the end user resulting from fuel properties outside the stated limits [31, 46, 47].

Density is the relationship between the mass of a liquid and its volume at a given temperature. Generally, density by itself does not provide much information about biodiesel, but complements the information provided by other properties.

Viscosity is a measure of the resistance of a substance to flow under the influence of gravity. Measuring viscosity in a fuel is crucial since it is a property that affects the operation of the fuel pump and the injection system of diesel engines, and exerts a significant influence on quality of the spray after injection in the cylinders. High viscosity can cause poor spray atomization, the formation of large drops and a high penetration of the fuel injection. This injection becomes a unified stream instead of a homogeneous mist of small droplets. As a result, the fuel is neither distributed nor mixed with the air to the required extent for a full burning, resulting in poor combustion and a loss of power and efficiency. On the other hand, low viscosities give a very thin spray and therefore less penetration. Again, this produces a poor combustion and a reduction in power and efficiency.

The flash point is defined as [32] the lowest temperature (corrected to a standard atmospheric pressure) in which the application of a test flame causes the vapour to ignite under specific trial conditions. A sample is considered to have achieved its flashpoint once a flame appears and immediately propagates across its surface. The flash point measures degree of response of a sample to heat and flame under controlled conditions. It is therefore a property that must be taken into account when assessing the risk of flammability of a fuel. It is used in transport and safety regulations to define flammable and combustible materials. It is important to note that biodiesel flashpoint must be greater than 100°C (values are usually higher than 160°C), while in petroleum diesel it should be above 50°C. This indicates that the biofuel has a significant advantage in safety issues for storage and transportation in comparison to its petroleum counterpart. Moreover, flashpoint is also used to determine the amount of alcohol remaining in the finished product (biodiesel), as explained in Chap. 8.

It is also important to determine the acid number of a fuel, since it is gives a measure of both inorganic and total acidity of the unused fuel and indicates a tendency to corrode metals in which it comes into contact. The acid number is also related to the to fuel degradation under service conditions. By definition, the acid number [41] represents the amount of base, expressed in milligrams of potassium hydroxide (KOH) per gram of sample, needed for the titration of a sample up to its end point.

The iodine value, however, is a measure of the degree of unsaturations in the fuel and it is directly related to the behavior of the same during storage condition since these unsaturations may lead to depositions. The value itself measures the amount of iodine needed to open (or break) double bonds of unsaturated organic compounds. This property depends on the origin of the feedstock, i.e., the vegetable oil from which the biofuel was produced. A high iodine value indicates a greater instability of the fuel.

It must be remarked that biodiesel is essentially sulfur-free [31].

Under low-temperature conditions (below 0°C), a wax precipitate may form in the fuel. This precipitate may block the lines and filters of fuel systems, causing malfunctions or even stopping the engines. The cloud point is defined as the temperature at which this wax precipitate begins to form. The purpose of this test is therefore to give an idea of the temperature at which the filter system will get blocked, restricting the fuel flow to the engine.

On the other hand, the pour point is a property that indicates the lowest temperature at which a fuel can be pumped. In other words, this parameter gives an idea of the lowest temperature at which a given fuel can be used.

Biodiesel typically has a higher pour point and cloud point than both the oils used as feedstock and petroleum diesel. This is important since it leads to difficulties for its use in cold climates, although this can be remedied by using additives.

Engine manufacturers recommend that the cloud point should be below the intended utilization temperature and no more than 6°C above the pour point.

The cetane number is also a very important indicator of fuel quality [38]. It is a measure of the behavior of a fuel in the engine and can only be experimentally determined using a special engine. The measured value represents the percentage of cetane (hexadecane, cetane number = 100) that must be mixed with heptamethylnonane (cetane number = 15) in order to have the same ignition performance of the fuel under test. It should be clearly understood, however, that the cetane scale is arbitrary and that there are substances with cetane numbers higher than 100. Its value relates the delay time to ignition with the injection in the combustion chamber. Higher values of the cetane number correspond to shorter ignition times. As a rule, a high cetane number indicates a quick and less noisy ignition.

The cetane number should not be confused with the cetane index. The cetane index is an alternative method used to obtain the cetane number through calculations, when the special engine required by ASTM D 613 is not available. These calculations are based on equations that relate values of other known properties of a fuel (such as gravity, API and average boiling point). The cetane index value calculated according to the formula is referred as the Calculated Cetane Index (ASTM D 976) [48]. Alternatively to the formula, a nomogram can be used. In general, however, calculations of the cetane index do not apply to non-conventional fuels (such as biodiesel).

When discussing the stability of biodiesel, it must be remembered that the term is quite broad and refers mainly to its thermal stability both at high and low temperatures, resistance to oxidation, polymerization, water absorption and microbial activity. Generally, the main sources of instability in biodiesel are the unsaturated fatty acid chains, although contact with metals and elastomers can also affect the stability during storage. The presence of water in the fuel can cause corrosion, especially when it is in contact with acids and hydroperoxides formed by the oxidation of biodiesel. Furthermore, the presence of water enhances microbial growth.

Biodiesel oxidation leads to the formation of hydroperoxides that can polymerize and form insoluble gums that may cause problems in the fuel system and filters.

The analytic method described in norm ASTM D 4530 [40] is used to determine the amount of residual carbon formed after the evaporation and pyrolysis of petroleum derivatives under certain conditions. The residual carbon is what remains after a sample undergoes thermal decomposition and depends on the specific conditions of evaporation and pyrolysis. The amount of residual carbon formed is an indication of the relative tendency of the tested substance to thermally degrade and form coke, under high-temperature pyrolytic conditions similar to those used in this analysis. It is an analytic method applied to those petroleum derivatives that partially decompose during distillation at atmospheric pressure. The "micro method" yields similar results as the ASTM D 189 [49] analytic method (Conradson residual carbon), with the advantage that it offers better control of the analytic conditions, smaller samples and requires less operator attention.

Characterization and quantification of linoleic acid methyl ester, ester content, free methanol, monoglycerides, diglycerides, triglycerides, free glycerin and total glycerin, is performed by chromatographic analysis. The Chap. 8 of this book presents an alternative procedure for the estimation of free methanol content, based on flash point and dielectric measurements.

A high level of mono-, di- and tri-glycerides indicates an insufficient conversion during the transesterification reaction, whereas a high content of free methanol indicates that the purification process was also inadequate. This high methanol content will originate a lower flash point value.

The free glycerol parameter indicates the presence of molecular glycerol in biodiesel, which can cause carbon deposits in the engine. A high content of free glycerol is due to poor separation of the ester–glycerol mixture phase as well as an inefficient purification process (washing). Higher values can originate deposits in the injectors as well as clogging in fuel systems. The free glycerin can separate from biodiesel and stay at the bottom of storage tanks and fuel systems.

The total glycerol parameter includes not only the free glycerol but also the remains of glycerol from non-reacted or partially reacted oils or fats (mono-, di- and tri-glycerides). Low levels ensure the high conversion of the reaction and purification of esters. High levels can cause injector deposits and may adversely affect cold weather operation and properties.

The copper strip corrosion test is used to determine the difficulties that may arise when the fuel comes into contact with the copper or bronze parts of the fuel system and cause potential corrosion.

Determining water and sediment content, as required by ASTM D 6751/09, is very important because it provides an indication of how clean the fuel is. Water is certainly undesirable since it can react with esters to form free fatty acids and it can increase microbial growth (as mentioned before) in storage tanks, reducing the maximum allowable storage time.

The sulfated ash test provides information about the residual ash left after the fuel is burned. This test is performed since there are materials that form ashes that can be originated in three ways: as abrasive solids, soluble metallic soaps and remnant catalyst. In the first and third cases they can affect injectors, fuel pump, pistons and form deposits in the engine. Soluble metallic soaps are less harmful but they can contribute to filter clogging and engine deposits.

In the case of biodiesel, the phosphorus comes from an incomplete refining process of the oil feedstock or from bones and proteins found in the rendering process of the oils.

Some properties depend on the feedstock, others from the actual chemical process and even others depend on both of them. For example, iodine value depends only on the oil, whereas methanol content, water content, acid number, phosphorous and esters content depends only on the process. Finally, viscosity and oxidation stability depend on both the oil and the process.

Consequently, there are oils such as soybean that, if used as feedstock for biodiesel production, will not yield a fuel that will comply with Pr EN 14214/09 since the iodine value will be close to 130, higher than the maximum established by the norm. The same will occur with low-temperature properties (cloud point, pour point, CFPP) of biodiesel from most animal fats, and also from oils with high percentage of saturated hydrocarbon chains (palm oil, coconut oil, etc.). Biodiesel produced from non-refined sunflower oil usually will not pass the total contamination test in EN 14214/03 and Pr EN 14214/09, whereas castor oil can have problems with the maximum limit of viscosity. Ester content of biodiesel from used vegetable oils will often be outside specified limits, indicating a low effi- ciency of the conversion reaction.

As it was mentioned in Chap. 2, the most suitable oils for biodiesel production are those with a high percentage of monounsaturated fatty acids.

To circumvent the limitations of the available feedstocks and achieve a final product that complies with the standards, two things can be done: either mix the available oil with another, more suitable oil, or simply mix the biodiesel produced from such oils with a biodiesel produced from feedstock with a high percentage of monounsaturated fatty acids, such as rapeseed. In both cases, a specific mixing percentage must be used so that the end product meets the standards.

The use of non-contaminating additives to adjust the values of some of these properties in order to comply with the standards is also permitted.

3.4 Characterization Properties not Required by Standards

There are other properties that are not yet incorporated within standards but are still very important for any kind of fuel.

In the first place, the power output of an engine while running under constant conditions and with a constant ratio of fuel consumption, is governed by its calorific value. A fuel with low calorific values yields less heat during combustion; therefore, the engine will provide less power than when operated with equal volume of a fuel with a higher calorific value. In order to maintain the power output when using a low calorific fuel, more fuel volume is needed. The calorific value of a fuel is defined as the amount of heat produced when the fuel is burnt completely. Essentially, the calorific value of biodiesel depends strictly on the calorific value of the vegetable oil used as feedstock. This relies heavily on the carbon number of the fatty acid chain (chain length) and the degree of unsaturation [50].

The measurement of the electrical properties (permittivity and dielectric loss) in the raw materials, in the different stages of the biodiesel manufacturing process and in the effluent (water used during washing steps), is also useful because it provides information about the reaction and of the different washing steps [51]. For example, the removal of contaminants in the oil after distillation is directly reflected in a reduction of dielectric losses. The measurement of permittivity and dielectric losses after each washing step clearly shows that it is possible to remove the last traces of methanol and sodium hydroxide. The same occurs with the measurements of electrical properties in the effluents, giving a quantitative indication of the efficiency of the washing and purification processes.

3.5 Quality Control of the Biodiesel Obtained in Different Production Scales

Ideally, all the fuel properties set by standards should be measured, to verify that they comply with the required values. This is the only way to ensure that the final product is indeed biodiesel. If the product meets the standards, there will be no engine problems.

This procedure is followed when the biodiesel is either for export or for domestic consumption. In plants that produce large volumes, quality controls are not a problem. Large-scale processes are continuous and the samples are analyzed in the plant laboratory, or periodically sent to external laboratories.

On the contrary, small producers cannot always afford the expenditure and personnel requirements involved in such controls. In plants where biodiesel production volumes are low, as in the case of self-consumption or small and medium companies, the actual process is usually done in batches and so the quality of the product may vary between batches. In these cases, having an in-house laboratory is economically unfeasible (the cost of most equipment is over several thousand dollars) and sending periodical samples to outside laboratories for analysis is also too costly. Therefore these small-scale producers tend to assume that the final product of the plant is actually biodiesel; this turns out to be an incorrect assumption in most cases. The problem of lack of quality controls in small and medium enterprises, including the farming and self-consumption sectors, is serious and global, and it is not just restricted to some countries.

3.6 Low-Cost Alternatives for Quality Control of Biodiesel

Some years ago, the Renewable Energy Group (GER) of the Engineering Faculty of the University of Buenos Aires (FI-UBA) faced budgetary limitations for the purchase of commercial equipment needed for characterization of biodiesel. Since one of the research lines of the Group include several aspects of quality control of biofuels, efforts were made in the design and construction of low-cost equipment and the development of chemical techniques to measure the properties required by standards. Some of the measuring systems that the Engineering group has designed and built include determination of the cloud point and pour point, copper strip corrosion, oxidation stability and flash point. The calibration was carried out with substances of known properties. Moreover, the results were checked by comparison with measurements on commercial equipment. All the designs were made according to the general guidelines indicated by applicable norms and standards and the construction was carried out with low-cost commercial parts. Also, procedures were put in practice for the determination of viscosity, density, acid value, iodine value, water content determination through Karl Fisher method and the calculation of free and total glycerol by an analytic method [52] alternative to the chromatographic procedure established by standards.

In addition, a research program was started to assess the usefulness of the measurement of electrical properties to quantify contaminants and provide information on the efficiency of each stage of the production process and on the quality of the raw materials and the final product. Although the measurement of electrical properties is not required at present by international standards, its successful application to biodiesel production opens exciting alternatives for quality control at small and medium enterprises and other technological applications. The automated pilot plant for the production of biodiesel designed and built by the GER at the FI-UBA will have an important role in the continuation of this research line.

3.7 Concluding Remarks

It is indisputable that performing the quality control of biodiesel is vital to ensure a satisfactory quality of the end product. Unfortunately, the complete quality control of the biofuel is usually carried out only in industries with large-scale production capacity. Most small and medium producers cannot always do so since the standards require the measurement of several properties and the necessary measuring equipment is simply unavailable. However, there is some equipment that can be designed and built at a reasonable cost, following the specifications set by the standards, that allows performing the measurement of some important parameters. On the other hand, measurement of electrical properties can also give relevant information about the quality of the biofuel. Both are useful alternatives for small- and medium-scale producers, as well as small research groups.

Acknowledgments The authors wish to thank Dr. Eng. C. Querini, Mr. M. Wainstein, BSc, MSc, and Mr. R. Marchetti, Lic. Econ., for their helpful comments and encouragement.

References

1. Meher LC, Vidya Sagar V, Naik SN (2006) Technical aspects of Biodiesel production by transesterification: a review. Renew Sustain Energy Rev 10(3):248–268
2. Pr EN 14214 (2009) Automotive fuels—fatty acid methyl esters (FAME) for diesel engines—requirements and test methods
3. EN 14103 (2003) Fat and oil derivatives—fatty acid methyl esters (FAME)—determination of ester and linolenic acid methyl ester contents
4. EN ISO 3675 (1998) Crude petroleum and liquid petroleum products—laboratory determination of density or relative density—hydrometer method
5. EN ISO 12185 (1996) Crude petroleum and petroleum products—determination of density— oscillating U-tube method
6. EN ISO 3104 (1996) Petroleum products—transparent and opaque liquids—determination of kinematic viscosity and calculation of dynamic viscosity
7. EN ISO 2719 (2002) Determination of flash point—Pensky-Martens closed cup method
8. EN ISO 3679 (2004) Determination of flash point—rapid equilibrium closed cup method
9. EN ISO 20846 (2004) Petroleum products—determination of sulfur content of automotive fuels—ultraviolet fluorescence method
10. EN ISO 20884 (2004) Petroleum products—determination of sulfur content of automotive fuels—wavelength-dispersive X-ray fluorescence spectroscopy
11. WI 019 376, Fatty acid methyl ester (FAME) as a blending component for diesel— determination of phosphorus and sulfur content—direct method by inductively coupled plasma optical emission spectrometric (ICP-OES)
12. EN ISO 5165 (1998) Diesel fuels—determination of ignition quality of diesel fuels—cetane engine method
13. ISO 3987 (1994) Petroleum products—lubricating oils and additives—determination of sulfated ash
14. EN ISO 12937 (2000) Petroleum products—determination of water—Coulometric Karl Fisher titration method
15. EN 12662 (2008) Liquid petroleum products—determination of contamination in middle distillates

16. EN ISO 2160 (1998) Petroleum products—corrosiveness to copper—copper strip test
17. EN 15751 (2009) Automotive fuels—fatty acid methyl ester (FAME) fuel and blends with diesel fuel—determination of oxidation stability by accelerated oxidation method
18. EN 14112 (2003) Fat and oil derivatives—fatty acid methyl esters (FAME)—determination of oxidation stability (accelerated oxidation test)
19. WI 019 370, Liquid petroleum products—fatty acid methyl ester (FAME) fuel and blends with middle distillates—determination of oxidation stability by rapid small scale oxidation method
20. EN 14104 (2003) Fat and oil derivatives—fatty acid methyl esters (FAME)—determination of acid value
21. EN 14111 (2003) Fat and oil derivatives—fatty acid methyl esters (FAME)—determination of iodine value
22. EN 15779 (2009) Petroleum products and fats and oil derivates—fatty acid methyl esters (FAME) for diesel engines—determination of polyunsaturated (≥ 4 double bonds) fatty acid methyl esters (PUFA) by gas chromatography
23. EN 14110 (2003) Fat and oil derivatives—fatty acid methyl esters (FAME)—determination of methanol content
24. EN 14105 (2003) Fat and oil derivatives—fatty acid methyl esters (FAME)—determination of free and total glycerol and mono-, di- and triglyceride content (reference method)
25. EN 14106 (2003) Fat and oil derivatives—fatty acid methyl esters (FAME)—determination of free glycerol content
26. EN 14108 (2003) Fat and oil derivatives—fatty acid methyl esters (FAME)—determination of sodium content by atomic absorption spectrometry
27. EN 14109 (2003) Fat and oil derivatives—fatty acid methyl esters (FAME)—determination of potassium content by atomic absorption spectrometry
28. EN 14538 (2006) Fat and oil derivatives—fatty acid methyl esters (FAME)—determination of Ca, K, Mg and Na content by optical emission spectral analysis with inductively coupled plasma (ICP OES) 2
29. EN 14107 (2003) Fat and oil derivatives—fatty acid methyl esters (FAME)—determination of phosphorus content by inductively coupled plasma (ICP) emission spectrometry
30. EN 116 (1997) Diesel and domestic heating fuels—determination of cold filter plugging point
31. ASTM D 6751 (2009) Standard specification for biodiesel fuel blend stock (B100) for middle distillate fuels
32. ASTM D 93 (2002) Test methods for flash-point by Pensky–Martens closed cup tester
33. ASTM D 2709 (2001) Test method for water and sediment in middle distillate fuels by centrifuge
34. ASTM D 445 (2003) Test method for kinematic viscosity of transparent and opaque liquids (the calculation of dynamic viscosity)
35. ASTM D 874 (2000) Test method for sulfated ash from lubricating oils and additives
36. ASTM D 5453 (2003) Test method for determination of total sulfur in light hydrocarbons, motor fuels, and oils by ultraviolet fluorescence
37. ASTM D 130 (2000) Test method for detection of copper corrosion from petroleum products by the copper strip tarnish test
38. ASTM D 613 (2003) Test Method for Cetane Number of Diesel Fuel Oil
39. ASTM D 2500 (2002) Test method for cloud point of petroleum products
40. ASTM D 4530 (2003) Test method for determination of carbon residue (micro method)
41. ASTM D 664 (2001) Test method for acid number of petroleum products by potentiometric titration
42. ASTM D 6584 (2000) Test Method for Determination of Free and Total Glycerine in B-100 Biodiesel Methyl Esters by Gas Chromatography
43. ASTM D 4951 (2002) Test method for determination of additive elements in lubricating oils by inductively coupled plasma atomic emission spectrometry
44. ASTM D 1160 (2003) Test method for distillation of petroleum products at reduced pressure

45. EN 14214 (2003) Automotive fuels—fatty acid methyl esters (FAME) for diesel engines—requirements and test methods
46. Van Gerpen J, Shanks B, Pruszko R, Clements D, Knothe G (2004) Biodiesel analytical methods. National renewable energy laboratory, NRRL/SR-510-36240
47. Van Gerpen J, Shanks B, Pruszko R, Clements D, Knothe G (2004) Biodiesel production technology. National Renewable Energy Laboratory, NRRL/SR-510-36244
48. ASTM D 976 (2000) Test methods for calculated cetane index of distillate fuels
49. ASTM D 189 (2001) Test method for Conradson carbon residue of petroleum products
50. Pryde EH (1983) Vegetable oils as diesel fuel: overview. J Am Oil Chem Soc 60:1557–1558
51. Sorichetti PA, Romano SD (2005) Physico-chemical and electrical properties for the production and characterization of biodiesel. Phys Chem Liq 43(1):37–48
52. Pisarello ML, Dalla Costa BO, Veizaga NS, Querini CA (2009) Volumetric method for free and total glycerine determination in biodiesel. Biomass Bioenergy (personal communication, submitted for publication)

Chapter 4
Electric Properties of Liquids

4.1 Introduction

This chapter presents an introduction to the study of the interaction of electromagnetic fields with liquids, and the parameters that characterize this interaction at the macroscopic level: permittivity and conductivity.

The description is based on the generalization of Maxwell classic field equations to linear, isotropic and homogeneous media for fields with harmonic time dependence. Approximations for different characteristic scales and the validity of the classical treatment are discussed.

Electromagnetic energy dissipation in materials and relaxation processes are described in terms of the frequency-dependent complex permittivity. The superposition of several relaxation processes is described phenomenologically in terms of the Havriliak–Negami dielectric function. A useful parameter for technological applications, the dissipation factor, is also defined.

The chapter closes with a discussion of polarization processes in liquids, particularly in connection to biodiesel production.

4.2 Electromagnetic Fields and Material Properties

4.2.1 Electromagnetic Fields in a Vacuum

Electromagnetic phenomena are described by Maxwell equations in terms of four vector fields: the electric field $\mathbf{E}$, the electric displacement $\mathbf{D}$, the magnetic induction $\mathbf{B}$ and the magnetic field $\mathbf{H}$. In a vacuum, $\mathbf{D} = \varepsilon_0 \, \mathbf{E}$ and $\mathbf{B} = \mu_0 \, \mathbf{H}$ [1].

In the international system of units (SI) that is used throughout this book, μ_0 ("permeability of free space") is, by definition, $4\pi \times 10^{-7}$ H/m and ε_0

S. D. Romano and P. A. Sorichetti, *Dielectric Spectroscopy in Biodiesel Production and Characterization*, Green Energy and Technology, DOI: 10.1007/978-1-84996-519-4_4, © Springer-Verlag London Limited 2011

("permittivity of free space") has the value of 8.8542×10^{-12} F/m. Therefore, the units of **E** and **D** are V/m and C/m^2, respectively; **H** is measured in A/m and **B** in T/m^2.

The electric field **E** and the magnetic field **B** determine the force **F** on a particle of electric charge q (measured in coulomb) moving with velocity **V** through the Lorentz law:

$$\mathbf{F} = q\mathbf{E} + \mathbf{V} \times \mathbf{B}. \tag{4.1}$$

Since there are no known point sources ("magnetic monopoles") for the magnetic field **B**, its divergence is zero:

$$\overline{\nabla} \cdot \mathbf{B} = 0. \tag{4.2}$$

On the contrary, free electric charges densities, ρ, and currents, **J**, are the sources of the fields **D** and **H**:

$$\overline{\nabla} \cdot \mathbf{D} = \rho \tag{4.3}$$

and

$$\overline{\nabla} \times \mathbf{H} = \mu_0 \mathbf{J} + \frac{\partial \mathbf{D}}{\partial t}. \tag{4.4}$$

Experimental evidence shows that electric charge is conserved, and therefore **J** and ρ are related by the equation of continuity:

$$\overline{\nabla} \cdot \mathbf{J} = \frac{\partial \rho}{\partial t} \tag{4.5}$$

Finally, **E** and **B** are related by the law of induction due to Faraday:

$$\overline{\nabla} \times \mathbf{E} = -\frac{\partial \mathbf{B}}{\partial t}. \tag{4.6}$$

The set of four differential equations 4.2–4.4 and 4.6 are known as Maxwell equations. Together with the Lorentz law, they describe electromagnetic fields and their interaction with charged particles in a vacuum, within the domain of applicability of classical approximations.

4.2.2 Characteristic Scales and Approximations

It must be remarked that ε_0 and μ_0 are not independent, since the speed of electromagnetic waves in a vacuum, the universal constant c (defined as exactly 2.997925×10^8 m/s) is related to ε_0 and μ_0 by:

$$c = \frac{1}{\sqrt{\mu_0 \varepsilon_0}}. \tag{4.7}$$

In the remainder of this book, unless explicitly stated, it will be assumed that the electromagnetic fields have a harmonic time dependence. In consequence, at a frequency f (i.e., angular frequency is $\omega = 2 \pi f$) the vacuum wavelength is:

$$\lambda = \frac{c}{f}. \tag{4.8}$$

Taking into account the characteristic size of the system to be described, L, in comparison with the wavelength, λ, different approximations are usually considered for the analysis of electromagnetic problems (Box 4.1).

Box 4.1 Characteristic Scales for Different Approximations

Network Theory $(L << \lambda)$. It deals with quasi-static treatment of the fields, based on potential theory. Kirchhoff's laws are applied to lumped circuit elements. Admittance and impedance are obtained from voltage and current measurements.

Transmission Line Formalism $(L \approx \lambda)$. Description of the fields is based on guided-wave modes. System parameters are defined and measured through reflection and transmission coefficients.

Geometrical Optics $(L >> \lambda)$. Fields are described in terms of free-space modes. Measurements are based on the amplitude of incident, transmitted and reflected waves in different directions.

It is clear from the above that specific experimental techniques must be used for dielectric measurements in each frequency range.

4.2.3 Classical Treatment of the Electromagnetic Field in Materials

On the molecular scale, the interaction between material particles and the electromagnetic field takes place by the exchange of photons of energy

$$E_p = hf \tag{4.9}$$

where h is Planck constant (experimentally, $h = 6.62618 \times 10^{-34}$ Js) and f is the frequency.

However, for a macroscopic system at an absolute temperature T, thermal fluctuations of molecular energies are of the order of:

$$\Delta E_T \approx k_B T. \tag{4.10}$$

Experimentally, $k_B = 1.3807 \times 10^{-23}$ J/K (Boltzmann constant).

To ascertain the validity of the classical treatment of the electromagnetic field in the interaction with macroscopic matter, it must be borne in mind that quantum effects will be relevant when the energy of individual photons becomes comparable to thermal fluctuations.

$$\Delta E_T \approx E_p. \tag{4.11}$$

In consequence, the classical description of the electromagnetic interaction is applicable if:

$$\frac{hf}{kT} \ll 1. \tag{4.12}$$

This condition is fulfilled in most dielectric measurements, including the microwave and millimeter wave ranges. For instance, for macroscopic systems near ambient temperature ($T \approx 300$ K) $\Delta E_T \approx 4.8 \times 10^{-17}$ J (26×10^{-3} eV) and the classical description of electromagnetic interactions will be valid for frequencies up to 10^{12} Hz, corresponding to infrared radiation.

4.2.4 Electromagnetic Fields in Materials

In material media, the presence of bound charges at the molecular level implies that the electric field **E** will differ from its value in a vacuum, $\mathbf{D}/\varepsilon_0$. Therefore, the effect of microscopic bound charges in materials may be described at the macroscopic level by the polarization vector **P** [2]:

$$\mathbf{P}(\omega)e^{-i\omega t} = \mathbf{D}(\omega)e^{-i\omega t} - \varepsilon_0\mathbf{E}(\omega)e^{-i\omega t}. \tag{4.13}$$

As stated in the previous section, it will be assumed that the electromagnetic fields have a harmonic time dependence and in consequence a complex exponential factor $e^{-i\omega t}$ (i.e., angular frequency $\omega = 2\pi f$) is implicit, unless indicated otherwise. It must be remarked that the complex representation is particularly useful since it simplifies notably the linear calculations with the field equations including derivatives with respect to the time variable. Of course, physical results are obtained by taking the real part of the complex values.

In linear, homogeneous and isotropic media, as will be the case for all the liquids studied in this book, **P** is proportional to **E**, and it may be written as:

$$\mathbf{P}(\omega) = \varepsilon_0\chi(\omega)\mathbf{E}(\omega). \tag{4.14}$$

The non-dimensional parameter $\chi(\omega) = \chi'(\omega) - i\,\chi''(\omega)$ is defined as the (complex) dielectric susceptibility of the media. Rearranging Eq. 4.13 and replacing with Eq. 4.14 results as:

$$\mathbf{D}(\omega) = \varepsilon_0(1 + \chi'(\omega) - i\chi''(\omega))\mathbf{E}(\omega). \tag{4.15}$$

In Eq. 4.6, it is convenient to define the complex permittivity of the material as

$$\varepsilon(\omega) = \varepsilon_0(1 + \chi(\omega)). \tag{4.16}$$

In consequence, the (non-dimensional) relative permittivity of the material is given by:

$$\varepsilon_r(\omega) = (1 + \chi(\omega)). \tag{4.17}$$

Therefore, in linear, isotropic and homogeneous materials, $\mathbf{D}(\omega)$ and $\mathbf{E}(\omega)$ are related by:

$$\mathbf{D}(\omega) = \varepsilon_0 \varepsilon_r(\omega) \mathbf{E}(\omega). \tag{4.18}$$

The complex relative permeability $\mu_r(\omega)$ is introduced to describe the effect of the material on magnetic fields, in a completely analogous way to the permittivity:

$$\mathbf{B}(\omega) = \mu_0 \mu_r(\omega) \mathbf{H}(\omega). \tag{4.19}$$

It must be remarked that in all the liquids studied in this book, the relative permittivity μ_r may be considered as real and equal to one. Finally, if there are mobile charge carriers in the material (for instance, when ionic substances are dissolved in a liquid), the current density $\mathbf{J}$ is proportional to the electric field $\mathbf{E}$ and to the conductivity $\sigma(\omega)$, usually considered as a real number:

$$\mathbf{J}(\omega) = \sigma(\omega)\mathbf{E}(\omega). \tag{4.20}$$

The Eqs 4.18–4.20 are often considered as the constitutive relations for linear, isotropic and homogeneous media. As summarized in Box 4.2 below, the complex relative permittivity $\varepsilon_r(\omega)$ and permeability $\mu_r(\omega)$, and the conductivity $\sigma(\omega)$ characterize the macroscopic response of the material to electromagnetic fields. Usually, these parameters are written as the sum of real and imaginary parts, (with the imaginary parts preceded by a minus sign). Plots of the real and complex parts of $\varepsilon_r(\omega)$ are usually referred to as dielectric spectra.

Box 4.2 Parameters of Linear, Isotropic and Homogeneous Materials

Complex relative permittivity. $\varepsilon_r(\omega) = \varepsilon'(\omega) - i\varepsilon''(\omega)$
Complex relative permeability. $\mu_r(\omega) = \mu'(\omega) - i\mu''(\omega)$
Conductivity. $\sigma(\omega)$

4.3 Electromagnetic Energy Dissipation in Materials

The instantaneous energy dissipation by unit time and unit volume (measured in W/m^3) originated by a current of mobile charge carriers $\mathbf{J}(t)$ in a material ("Joule effect") is given by

$$w_J(t) = \mathbf{J}(t) \cdot \mathbf{E}(t) \tag{4.21}$$

(where the time dependence of the fields is arbitrary). This may be considered as the work done (by unit time) by the electric field on the moving charges as follows from Lorentz law, equation Eq. 4.1. For fields with harmonic time dependence, it is easy to show that the average in one period of the scalar product in Eq. 4.21 may be obtained as:

$$\langle \mathbf{J}(t) \cdot \mathbf{E}(t) \rangle = \frac{1}{2} \mathrm{Re}[\mathbf{J}^*(\omega) \cdot \mathbf{E}\omega]. \tag{4.22}$$

where $\mathbf{J}^*(\omega)$ is the complex conjugate of the complex representation of $\mathbf{J}(t)$. In consequence, from the constitutive relation 4.20, in linear, isotropic and homogeneous materials the average in one period of the dissipation per unit volume is proportional to the square of the amplitude of the electric field, $\mathbf{E}$:

$$\langle w_J \rangle = \frac{1}{2} \mathrm{Re}[\sigma \mathbf{E}^*(\omega) \cdot \mathbf{E}(\omega)] = \frac{1}{2} \sigma |\mathbf{E}|^2. \tag{4.23}$$

The imaginary part of the complex permittivity parameter, defined in the previous section, represents additional electromagnetic energy losses due to dissipative processes at the molecular level, associated with bound charges. The polarization response of the material is delayed with respect to the excitation and, in consequence, the macroscopic polarization $\mathbf{P}$ must be written as the sum of a component in phase with $\mathbf{E}$, associated with the real part of the susceptibility and another at a phase angle of $\pi/2$, related to its imaginary part, as follows from Eq. 4.14. From Eq. 4.15 it is clear that the electric displacement vector $\mathbf{D}$ will also have a component in phase with $\mathbf{E}$, proportional to $(1 + \varepsilon'(\omega))$, and another lagging in $\pi/2$ proportional to the imaginary part $\varepsilon''(\omega)$.

To clarify the physical meaning of both components of $\mathbf{D}$, it is useful to consider a small plane parallel capacitor in a vacuum. In the quasi-static approach, the free charge on the plates is given by:

$$Q(\omega) = C_0 V(\omega). \tag{4.24}$$

It must be remarked that, from Eq. 4.3 and applying Gauss theorem, the magnitude of the electric displacement vector $\mathbf{D}$ is the charge per unit area, Q/A. Similarly, the potential difference V divided by the distance between the capacitor plates, V/d, gives the magnitude of the electric field $\mathbf{E}$. The capacitance C_0 is then proportional to the ratio between the magnitudes of $\mathbf{D}$ and $\mathbf{E}$, and therefore to the vacuum permittivity ε_0. Neglecting fringe effects, C_0 is proportional to the area of the plates A, and inversely proportional to the distance between the plates, d:

$$C_0 = \varepsilon_0 \frac{A}{d}. \tag{4.25}$$

The current is given by the time derivative of the charge:

$$I_c = \frac{dQ}{dt} \tag{4.26}$$

and in complex notation it may be written as:

$$I_c(\omega) = i\omega C_0 V(\omega). \tag{4.27}$$

It may be seen that there is a phase angle of $\pi/2$ between the current and the voltage, since $\mathbf{D}$ and $\mathbf{E}$ are in phase and in consequence there is no power dissipation.

If the capacitor is filled with a dielectric of complex relative permittivity $\varepsilon_r(\omega)$, $\mathbf{D}$ and $\mathbf{E}$ are no longer in phase. Therefore, the capacitance C_0 changes to a complex value $C = \varepsilon_r(\omega) C_0$ and the current I has a component in phase with the voltage V that is proportional to $\varepsilon''(\omega)$:

$$I_C(\omega) = (\omega\varepsilon''(\omega) + i\omega\varepsilon'(\omega))C_0 V(\omega). \tag{4.28}$$

The in-phase component of the current originates a power dissipation that is proportional to $\varepsilon''(\omega)$, as distinct from the component proportional to $\varepsilon'(\omega)$ related to the energy stored in the dielectric.

If in the material there are also mobile charge carriers, an additional conduction current term must be added, proportional to the conductivity σ:

$$I_J(\omega) = \sigma\frac{A}{d}V(\omega) \tag{4.29}$$

and the total current $I_C = I_J + I_S$ is:

$$I_C(\omega) = \left(\omega\varepsilon''(\omega) + \frac{\sigma}{\varepsilon_0} + i\omega\varepsilon'(\omega)\right)C_0 V(\omega). \tag{4.30}$$

The total power dissipation in the capacitor, W_T, averaged during one period, is given by the product of the real parts of $I_C(\omega)$ and $V(\omega)$:

$$\langle W_T \rangle = \frac{1}{2}\mathrm{Re}[I_C(\omega)]\mathrm{Re}[V(\omega)]. \tag{4.31}$$

Therefore,

$$\langle W_T \rangle = \frac{1}{2}\left(\omega\varepsilon''(\omega) + \frac{\sigma}{\varepsilon_0}\right)C_0(V(\omega))^2. \tag{4.32}$$

Since in the plane parallel capacitor $V = d\,\mathbf{E}$ and $C_0 = \varepsilon_0 A/d$, the total average power dissipated per unit volume is:

$$\langle w_T \rangle = \frac{1}{2}(\omega\varepsilon_0\varepsilon''(\omega) + \sigma)|\mathbf{E}|^2 \tag{4.33}$$

where the total dissipation w_T may be written as the sum of a term due to mobile charge carriers, w_J (also called "conduction losses"), and another due to bound charges, w_D ("dielectric losses"):

$$\langle w_T \rangle = \langle w_D \rangle + \langle w_J \rangle. \tag{4.34}$$

Comparing with Eq. 4.23, w_D is given by:

$$\langle w_D \rangle = \frac{1}{2} \omega \varepsilon_0 \varepsilon''(\omega) |\mathbf{E}|^2. \tag{4.35}$$

The previous relations are summarized in Box 4.3.

Box 4.3 Power Dissipation (per unit volume) in Linear and Isotropic Materials Under Time-Harmonic Excitation

Total power dissipation. $\langle w_T \rangle = \langle w_D \rangle + \langle w_J \rangle$ (averaged in one period).
Dielectric losses. $\langle w_D \rangle = \frac{1}{2} \omega \varepsilon_0 \varepsilon''(\omega) |\mathbf{E}|^2$.
Conduction losses. $\langle w_J \rangle = \frac{1}{2} \sigma |\mathbf{E}|^2$.

From the above, it is clear that dissipative processes due to bound and mobile charges may be grouped together in the imaginary part of the complex relative permittivity:

$$\varepsilon_r(\omega) = \varepsilon'(\omega) - i \left[\varepsilon''(\omega) + \frac{\sigma}{\varepsilon_0 \omega} \right]. \tag{4.36}$$

It is also interesting to remark that analogous expressions may be written for magnetic materials, where losses are proportional to the imaginary part of the relative permeability, μ''. However, for all the liquids studied in this book, magnetic losses are negligible.

At infrared and optical frequencies, where the description is based on wave propagation (cf. Sect. 4.2.2), dielectric losses are measured from wave attenuation. For non-magnetic liquids ($\mu_r = 1$), the complex refractive index $n(\omega)$ is defined as:

$$n(\omega) = \sqrt{\varepsilon'(\omega) - i\varepsilon''(\omega)} \tag{4.37}$$

and the complex wave number $k(\omega) = k'(\omega) - i\, k''(\omega)$ is given by

$$k(\omega) = \frac{\omega}{c} n(\omega). \tag{4.38}$$

The electric field of an electromagnetic wave with harmonic time dependence may be written as:

$$\mathbf{E}(x,t) = \mathbf{E}_0 e^{-ik(\omega)x} e^{i\omega t}. \tag{4.39}$$

Separating the real and imaginary parts of $k(\omega)$ and rearranging it results as:

$$\mathbf{E}(x,t) = \mathbf{E}_0 e^{-k''(\omega)x} e^{-ik'(\omega)x} e^{i\omega t}. \tag{4.40}$$

Therefore, Eq. 4.39 with a complex wavenumber describes a wave that is attenuated exponentially when propagating in medium with dielectric losses:

$$\mathbf{E}(x,t) = \mathbf{E}_0 e^{-\frac{2\pi x}{\delta}} e^{-i\left(\frac{2\pi}{\lambda}x - \omega t\right)}. \tag{4.41}$$

It is easy to see that the real part $k'(\omega)$ determines the wavelength, λ:

$$k'(\omega) = \frac{2\pi}{\lambda(\omega)} \tag{4.42}$$

and the complex part $k''(\omega)$ sets the characteristic length of attenuation, δ:

$$k''(\omega) = \frac{2\pi}{\delta(\omega)}. \tag{4.43}$$

For liquids with low dielectric losses, $\varepsilon''(\omega) << \varepsilon'(\omega)$ and in consequence:

$$\sqrt{\varepsilon'(\omega) - i\varepsilon''(\omega)} \approx \sqrt{\varepsilon'(\omega)} - \frac{i}{2}\sqrt{\varepsilon''(\omega)}. \tag{4.44}$$

With this approximation, the real and imaginary parts of $k(\omega)$ are:

$$k'(\omega) = \frac{\omega}{c}\sqrt{\varepsilon'(\omega)} \tag{4.45}$$

and

$$k''(\omega) = \frac{1}{2}\frac{\omega}{c}\sqrt{\varepsilon''(\omega)}. \tag{4.46}$$

It follows that the wavelength λ and attenuation length δ are given by:

$$\lambda(\omega) = 2\pi \frac{c}{\omega} \frac{1}{\sqrt{\varepsilon'(\omega)}} \tag{4.47}$$

and

$$\delta(\omega) = 4\pi \frac{c}{\omega} \frac{1}{\sqrt{\varepsilon''(\omega)}}. \tag{4.48}$$

In the infrared and optical range, measurements of the refractive index usually refer to the real part only. Since these measurements are made at wavelengths where the liquid is transparent (outside absorption bands), the low-loss approximation is applicable. In consequence, the real part 4.45 is related to the real part of permittivity through Eq. 4.45. On the other hand, attenuation measurements give the imaginary part of the permittivity through Eqs. 4.37, 4.38 and 4.43.

It must be remembered that if the attenuation is strong, for instance in the absorption bands in the infrared range, the low-loss approximation (4.44) no longer holds and Eqs. 4.47 and 4.48 must be written in terms of the real and imaginary parts of the complex refractive index:

$$\lambda(\omega) = 2\pi \frac{c}{\omega} \frac{i}{\mathrm{Re}\sqrt{\varepsilon'(\omega) - i\varepsilon''(\omega)}} \tag{4.49}$$

$$\delta(\omega) = 2\pi \frac{c}{\omega} \frac{1}{\mathrm{Im}\sqrt{\varepsilon'(\omega) - i\varepsilon''(\omega)}} \tag{4.50}$$

It is important to remark that refractive index measurements in the visible range are useful for the characterization of vegetable oils and also mixtures of fatty acid esters, as it will be shown in Chaps. 6 and 7.

4.4 Dielectric Relaxation

4.4.1 The Debye Model: Polarization Relaxation with a Single Characteristic Time

The frequency dependence of the complex permittivity is intimately related to the characteristic time scales of the relaxation processes of molecular polarization. If a perturbed dielectric sample with initial polarization $\mathbf{P}(0)$ returns to its equilibrium value $\mathbf{P}_{eq}$ by a relaxation process of characteristic time τ_0, to the first order, the process may be described by a linear differential equation:

$$\frac{d\mathbf{P}(t)}{dt} = \frac{\mathbf{P}(0) - \mathbf{P}_{eq}}{\tau_0}. \tag{4.51}$$

Therefore, the time evolution of the macroscopic polarization is given by:

$$\mathbf{P}(t) = \left(\mathbf{P}(0) - \mathbf{P}_{eq}\right)e^{-\frac{t}{\tau_0}} + \mathbf{P}_{eq}. \tag{4.52}$$

If, as in previous sections, a harmonic time dependence is assumed for the electric field $\mathbf{E}$, the polarization $\mathbf{P}(t)$ will also be harmonic. In consequence, the steady-state solution of Eq. 4.51 at an angular frequency ω is given by:

$$\mathbf{P}(\omega) = \mathbf{P}_\infty + \frac{\mathbf{P}_o - \mathbf{P}_\infty}{1 + i\omega\tau_0} \tag{4.53}$$

where $\mathbf{P}_\infty$ and $\mathbf{P}_0$ are the asymptotic values of $\mathbf{P}(\omega)$ in the limits of very high frequencies ($\omega \to \infty$) and very low frequencies ($\omega \to 0$), respectively.

From Eq. 4.14, for linear, isotropic and homogeneous media, the complex susceptibility for a dielectric with a relaxation process with a characteristic time τ_0 may be written as:

$$\chi'(\omega) - i\chi''(\omega) = \chi_\infty + \frac{\chi_o - \chi_\infty}{1 + i\omega\tau_0} \tag{4.54}$$

and from Eqs. 4.15–4.17, the complex relative permittivity is

$$\varepsilon'(\omega) - i\varepsilon''_{\text{pol}}(\omega) = \varepsilon_\infty + \frac{\varepsilon_o - \varepsilon_\infty}{1 + i\omega\tau_0}. \tag{4.55}$$

In Eqs. 4.54 and 4.55, the sub-indexes 0 and ∞ indicate the asymptotic values of the parameters in the limits of very high frequencies ($\omega \to \infty$) and very low frequencies ($\omega \to 0$), respectively. Dielectric processes with single relaxation times were first modeled by Debye, and Eq. 4.55 is known as the Debye dielectric function.

The real part of 4.55 tends to constant asymptotic values for very high and very low frequencies. The difference between the asymptotic values is termed the "relaxation intensity":

$$\Delta\varepsilon = \varepsilon_o - \varepsilon_\infty. \tag{4.56}$$

Therefore, $\varepsilon'(\omega)$ decreases gradually around the characteristic frequency $1/\tau_0$.

$$\varepsilon'(\omega) = \varepsilon_\infty + \frac{\Delta\varepsilon}{1 + (i\omega\tau_0)^2} \tag{4.57}$$

.

Similarly, when plotted in a logarithmic scale, $\varepsilon''(\omega)$ has a broad symmetric peak centered at $\omega = 1/\tau_0$:

$$\varepsilon''(\omega) = \frac{\omega\Delta\varepsilon}{1 + (i\omega\tau_0)^2}. \tag{4.58}$$

It is easy to see that the dissipation in the material reaches its maximum value when the angular frequency of the field is the inverse of the relaxation time τ_0, since the phase lag between **D** and **E** is greatest. At frequencies much lower or much higher than $1/\tau_0$, the phase lag tends to zero and there is no dielectric loss in the material.

4.4.2 Dielectrics with Several Relaxation Times

If several relaxation processes are present in a dielectric, Eq. 4.38 may be generalized as a sum of Debye processes:

$$\varepsilon'(\omega) - i\varepsilon''_{\text{pol}}(\omega) = \varepsilon_\infty + \sum_n \frac{\Delta\varepsilon_n}{1 + i\omega\tau_n} \tag{4.59}$$

where $\Delta\varepsilon_n$ is the relaxation intensity of the process with characteristic time τ_n. This expression is useful to describe the dielectric properties of materials when the relaxation processes have very different times. Therefore, the peaks in $\varepsilon''(\omega)$ are well separated and each one may be fitted to a Debye process.

4.4.3 Superposition of Multiple Relaxation Times: the Havriliak–Negami Dielectric Function

In some dielectrics, it is found the experimental data cannot be accurately fitted to a Debye function. For instance, the peaks in $\varepsilon''(\omega)$ are broader and non-symmetric. This behavior is explained by assuming that each peak results from the superposition of many Debye processes with very close relaxation times. In this case, it is very difficult to fit the observed values of permittivity to the general form given by (4.59), since it can be shown that the mathematical problem is "ill-posed", that is, the fitted values vary widely for very small changes in the experimental data. In consequence, to characterize these processes, it is preferable to a phenomenological model proposed by Havriliak and Negami [3], as a generalization of the Debye dielectric function:

$$\varepsilon'(\omega) - i\varepsilon''_{pol}(\omega) = \varepsilon_\infty + \frac{\varepsilon_o - \varepsilon_\infty}{\left[1 + (i\omega\tau_0)^\alpha\right]^\beta}. \tag{4.60}$$

The two additional parameters α and β describe the broadening of the relaxation peak with respect to the Debye model (that corresponds to $\alpha = \beta = 1$). The parameter α describes the symmetrical and β the asymmetrical broadening, respectively. Both parameters must be positive and the product $\alpha\beta$ must be less than 1. If $\beta = 1$, the process is of the Cole–Cole type, and for $\alpha = 1$, the model is called the Cole–Davidson function.

4.5 Dielectric Polarization in Liquids

Several processes at the molecular scale, with different characteristic times, contribute to dielectric polarization in liquids, as summarized in Box 4.4. The measurement techniques for each frequency range are presented in Chap. 5.

4.5.1 Electronic Polarization

This process is originated by the dipole moment induced by the interaction of the electric field with the electrons in the molecule. Since the characteristic relaxation times are of the order of 10^{-11} s or less, this process is dominant at the highest frequencies, typically above 10^{10} Hz (millimeter waves). It controls the refractive index of liquids and originates absorption bands at infrared and optical frequencies.

In the absence of contaminants, electronic polarization is the only observable relaxation process in non-polar and non-conductive liquids such as biodiesel and pure vegetable oils. In the context of biodiesel production and characterization, it is important to remark that infrared absorption bands in the micrometer range are very useful for the identification of methyl esters and triglycerides. Quantitative assessment of these substances is of critical interest for production and quality control, since it provides a measure of the efficiency of the conversion process.

At optical frequencies, above the electronic relaxation bands, the refractive index of biodiesel, fatty acid methyl esters (FAME) mixtures and vegetable oils, to be discussed in Chap. 5, although numerically close, can be readily distinguished by refractometric techniques [5].

4.5.2 Orientation Polarization

Molecules that have a permanent dipole moment, such as water and alcohols, tend to orient in the direction of the field. The characteristic relaxation times for this relaxation process are in the range from 10^{-8} to 10^{-10} s, depending on the molecular structure and temperature of the sample. In general, the relaxation intensity diminishes with increasing temperatures and is stronger in small, more easily oriented molecules.

This process controls the permittivity in the range from about 10^{8} Hz up to 10^{10} Hz (very high frequencies, VHF, to microwave). Methanol, one of the raw materials for biodiesel production, has an orientation relaxation process at a frequency of about 4 GHz, whereas water has one near 20 GHz.

4.5.3 Interfacial Polarization

A dipole layer appears between two phases of different permittivity and conductivity, or between metallic electrodes and a dielectric with non-negligible conductivity. This effect was initially studied by Maxwell in the nineteenth century and plays an important role in many low-frequency dielectric phenomena in liquids. It is termed Maxwell–Wagner or interfacial polarization [4] and originates a strong relaxation process at low frequencies, with characteristic times below 10^{-7} s. Particularly, interfacial polarization is an important effect in the measurement of permittivity of conductive liquids below 10^{8} Hz, down to sub-Hertz frequencies, appearing as an increase of the real part of permittivity at low frequencies. For instance, dielectric measurements on the effluents of biodiesel washing steps [5], to be discussed in Chap. 6, show strong interfacial polarization effects.

Box 4.4 Polarization Processes in Liquids

Electronic polarization.
Characteristic relaxation times are of the order of 10^{-11} s or less.

It is the dominant process in non-polar liquids, controls the refractive index, and originates absorption bands at infrared and optical frequencies.
Orientation polarization.
The characteristic relaxation times are in the range from 10^{-8} to 10^{-10} s.

This process controls the permittivity of polar molecules in the range from about 10^8 up to 10^{10} Hz.
Interfacial polarization.
The characteristic relaxation times are below 10^{-7} s. Interfacial polarization controls the permittivity below 10^8 Hz, down to sub-Hertz frequencies, particularly in conductive liquids in contact with metallic electrodes.

4.6 Concluding Remarks

Within the domain of the classical approach presented in this chapter, the interaction of electromagnetic fields with materials is characterized by three macroscopic parameters: permittivity, permeability and conductivity.

In non-magnetic, linear, isotropic and homogeneous media, such as all the liquids studied in this book, the dielectric behavior under time-harmonic excitation is described by the complex permittivity. This characterizes at the macroscopic scale the polarization processes and dissipative effects in the liquid.

The characteristic times of the different dielectric polarization processes, electronic, orientation and interfacial, determine the frequency dependence of the complex permittivity and are related to the properties of the liquid at the molecular scale, as described in the following chapters.

References

1. Jackson JD (1975) Multipoles, electrostatics of macroscopic media and dielectrics. In: Classical electrodynamics, 2nd edn. Wiley, Inc., New York
2. von Hippel AR (1954) Macroscopic approach. In: Dielectrics and waves, 1st edn. Wiley, Inc., New York
3. Havriliak S, Negami S (1967) A complex plane representation of dielectric and mechanical relaxation processes in some polymers. Polymer 8(4):161–210

4. Coelho R, Aladenize B (1993) Polarisation, permittivité et relaxation. In: Les Dielectriques, 1st edn. Hermes, Paris
5. Sorichetti PA, Romano SD (2005) Physico–chemical and electrical properties for the production and characterization of biodiesel. Phys Chem Liq 43(1):37–48

Chapter 5
Introduction to Dielectric Spectroscopy

5.1 Introduction

Dielectric spectroscopy (DS) is based on the interaction of a macroscopic sample with a time-dependent electric field.

In the lower frequency range, where network theory is adequate, the substance under study is placed in a cell with a system of electrodes connected to the measuring circuit. A generator excites the cell and the response signal is compared with the excitation signal to determine the dielectric properties of the sample, usually through admittance measurements. This approach is adequate when the size of the cell and connection cables is much shorter than the wavelength of the excitation signal. In most measurement systems, this corresponds to a maximum frequency of about 20 MHz.

For dielectric measurements at higher frequencies a description based on transmission line theory becomes necessary, and dielectric properties are obtained from measurements the reflection coefficient of the cell containing the sample, placed at the end of a coaxial transmission line (or a waveguide, at microwave and millimeter wave frequencies).

Since the dielectric properties of liquids depend strongly on temperature, the cell is usually placed in a thermostat during measurements and several isothermal dielectric spectra are obtained at different temperatures. Relevant information on liquid samples properties can be obtained from the analysis of the dependence of the parameters of dielectric spectra on temperature.

5.2 Measurement of Dielectric Properties at Low Frequencies

Measurement of dielectric properties at low frequencies, where the circuit theory approach is applicable (usually below 20 MHz), is based on the determination of the admittance of the cell containing the sample, as a function of frequency [1–4].

S. D. Romano and P. A. Sorichetti, *Dielectric Spectroscopy in Biodiesel Production and Characterization*, Green Energy and Technology, DOI: 10.1007/978-1-84996-519-4_5,
© Springer-Verlag London Limited 2011

5.2.1 Admittance, Capacitance, and Permittivity

Precision dielectric measurements in the low-frequency range are usually made by comparing the admittance (or impedance) of the sample cell with a reference capacitor.

The capacitance of the empty cell, C_0, is

$$C_0 = K_0 \varepsilon_0 \tag{5.1}$$

where ε_0 is the free space permittivity and K_0 is the "cell constant" which depends on the geometry of the cell. Therefore, the admittance of the empty cell under sinusoidal excitation at a frequency f, neglecting losses and inductive effects, corresponds to a purely imaginary (capacitive) susceptance:

$$Y_0(\omega) = i\omega C_0 \tag{5.2}$$

It must be noted that inductive effects can be neglected if the length of the transmission lines is much shorter than the wavelength at the highest measurement frequency, as is usually the case for most measurement systems for frequencies up to 30 MHz.

When the measurement cell is filled with the sample, the measured value of the admittance at the angular frequency ω takes the complex value $Y_s(\omega)$, that may be written as the sum of a real conductance $G(\omega)$, originated in dissipative processes in the sample, and a an imaginary (capacitive) susceptance $B(\omega)$ related to dielectric polarization:

$$Y_s(\omega) = G(\omega) + iB(\omega) \tag{5.3}$$

It is worth noting that, since the conductance, $G(\omega)$, corresponds to energy dissipation in the sample, it must be non-negative (the limiting case of a sample without losses corresponds to a purely imaginary admittance).

At this point it is convenient to generalize Eq. 5.3 by introducing the equivalent complex capacitance of the cell, $C(\omega)$. This is a complex number such that

$$Y_s(\omega) = i\omega C(\omega). \tag{5.4}$$

Usually, the imaginary part of $C(\omega)$ is written with a minus sign:

$$C(\omega) = C'(\omega) - C''(\omega) \tag{5.5}$$

This leads to a non-negative value of the real part of the admittance for positive values of $C''(\omega)$, as it is easy to see from Eq. 5.4:

$$Y_s(\omega) = i\omega[C'(\omega) - iC''(\omega)]. \tag{5.6}$$

In summary, the real and imaginary parts of the complex capacitance $C(\omega)$ are directly related to the measured values of conductance and susceptance:

$$\omega C'(\omega) = B(\omega) \tag{5.7}$$

$$\omega C''(\omega) = G(\omega) \tag{5.8}$$

Finally, the apparent complex relative permittivity of the sample, $\varepsilon_r(\omega)$, at an angular frequency ω is defined as the ratio between the complex capacitance of the cell with the sample, $C(\omega)$, and the capacitance of the empty cell, C_0:

$$\varepsilon_r(\omega) = \frac{C(\omega)}{C_0} \tag{5.9}$$

where the relative complex permittivity is usually written as:

$$\varepsilon_r(\omega) = \varepsilon'(\omega) - i\varepsilon''(\omega) \tag{5.10}$$

If the sample, in addition to dielectric polarization, has charge transport effects, these are customarily included in the imaginary part of the complex relative permittivity. Therefore, $\varepsilon''(\omega)$ is written as the sum of a term related to the dielectric polarization, $\varepsilon''_{pol}(\omega)$, and a conductivity term $\sigma/(\omega\varepsilon_0)$,

$$\varepsilon_r(\omega) = \varepsilon'(\omega) - i\varepsilon''_{pol}(\omega) - i\frac{\sigma}{\omega\varepsilon_0} \tag{5.11}$$

The term $\varepsilon''_{pol}(\omega)$ describes the dissipation of energy associated to the relaxation processes of dielectric polarization, and it is related to the real part $\varepsilon'(\omega)$ through the Kramers–Kronig relations. On the other hand, the conductivity terms $\sigma/(\omega\varepsilon_0)$ describes the dissipation associated to charge transport phenomena, and at low frequencies $(\omega \to 0)$ dominates the imaginary part of the complex permittivity.

5.2.2 Determination of the Cell Constant

The cell constant K_0, or, equivalently, the capacitance of the empty cell, C_0, is obtained from the difference in capacitance values of the empty cell and when filled with a liquid of known relative permittivity.

Typical reference liquids for the calibration of sample cells include cyclohexane, carbon tetrachloride and methanol.

It is very important to minimize the uncertainty on the value of C_0, since it will have a direct impact on all dielectric measurements. Therefore, the guidelines presented in Sect. 5.4 must be strictly followed.

The value of the cell constant must be checked at periodic intervals, or when any kind of damage to the cell is suspected. Moreover, since the capacitance of the empty cell must be accurately known, the presence of contaminants due to an inadequate cleaning can be readily detected by comparison of the measured capacitance of the cell, as indicated in Sect. 5.4.3.

5.2.3 Dissipation Factor

The dissipation factor of the sample, $D(\omega)$, is defined as the tangent of the phase angle $\delta(\omega)$ between the excitation voltage and the current through the cell. Therefore, the dissipation factor is given by the quotient (with a minus sign) between the imaginary and real parts of $\varepsilon_r(\omega)$:

$$D(\omega) = \frac{\varepsilon''(\omega)}{\varepsilon'(\omega)}. \tag{5.12}$$

The dissipation factor may be understood as the ratio between the energy dissipated (proportional to $\varepsilon''(\omega)$) and the energy stored (proportional to $\varepsilon'(\omega)$) in the dielectric, per unit cycle and per unit volume. Therefore, it is a useful "figure of merit" for technological applications, particularly in electric power applications. Results of complex permittivity measurements are usually given as plots of the real part of permittivity, $\varepsilon'(\omega)$ and the imaginary part $\varepsilon''(\omega)$. In technological applications, plots of the dissipation factor $D(\omega)$ are also frequently given.

In conductive liquids, the dissipation factor $D(\omega)$ varies as ω^{-1} in the low-frequency range, where the conductivity term dominates dielectric losses. This is apparent in a logarithmic plot of $D(\omega)$.

5.3 Measurement of Dielectric Properties at Higher Frequencies

As indicated in Sect. 4.2.2, when the characteristic dimensions of the sample cell and connection cables become comparable to the wavelength of the excitation field, the quasi-static treatment of fields, based on potentials and Kirchoff's equations applied to lumped circuit elements is no longer adequate. The use of transmission line concepts becomes necessary and system parameters are defined in terms of reflection and transmission coefficients [1–4].

5.3.1 Propagation in Coaxial Transmission Lines

For measurement systems up to several GHz, coaxial lines are frequently used. Within the validity of the transmission line formalism, it is assumed that propagation in coaxial lines takes place in the transverse electromagnetic (TEM) mode. In the TEM mode, the electric field lies in the plane perpendicular to the line axis, and it is possible to define the voltage between the conductors of the line and also the current flowing through them. Therefore, electromagnetic fields propagation in the line is described in terms of current and voltage wave solutions.

On a plane located at a distance x from a reference plane, the instantaneous voltage $V(x, t)$ is written as the sum of an incident wave $V_{inc}(x, t)$ and a reflected voltage wave, $V_{inc}(x, t)$:

$$V(x,t) = V_{inc}(x,t) + V_{ref}(x,t). \tag{5.13}$$

Similarly, the instantaneous current through the conductors at x, $I(x, t)$ is the superposition of the incident and reflected current waves, $I_{inc}(x, t)$ and $I_{ref}(x, t)$:

$$I(x,t) = I_{inc}(x,t) + I_{ref}(x,t). \tag{5.14}$$

The incident and reflected current and voltage waves are related through the characteristic impedance of the line, Z_0:

$$\frac{V_{inc}(x,t)}{I_{inc}(x,t)} = Z_0 \tag{5.15}$$

$$\frac{V_{ref}(x,t)}{I_{ref}(x,t)} = Z_0. \tag{5.16}$$

Most transmission lines for measurement applications are, by construction, homogeneous and with very low losses. Therefore, Z_0 is usually assumed to be real and constant.

5.3.2 Transmission Lines Excited by Harmonic Signals

Assuming that the excitation is harmonic, Eq. 5.13 may be written as:

$$V(\omega,x) = U_{inc}e^{i\omega t}e^{-ikx} + U_{ref}e^{i\omega t}e^{ikx} \tag{5.17}$$

then, if the phase velocity of the waves in the line is v, the wavelength in the line, λ, and the wavenumber, k, are given by:

$$\lambda = \frac{2\pi v}{\omega} \tag{5.18}$$

$$k = \frac{\omega}{v}. \tag{5.19}$$

It is important to remark that if the transmission line may be considered as ideal, the complex amplitudes U_{inc} and U_{ref} are constant along the line.

As indicated in the previous chapter, the time exponential factor $e^{i\omega t}$ will be implicit since a harmonic time dependence will be assumed unless otherwise indicated.

5.3.3 Reflection Coefficient, Impedance and Admittance

The reflection coefficient at the plane located at x, $\rho(\omega, x)$, is defined as:

$$\rho(\omega, x) = \frac{V_{\text{ref}}(\omega, x)}{V_{\text{inc}}(\omega, x)} \quad (5.20)$$

furthermore, introducing in Eq. 5.20 the complex amplitudes U_{inc} and U_{ref} it results:

$$\rho(\omega, x) = \frac{U_{\text{ref}}(\omega, x)}{U_{\text{inc}}(\omega, x)} e^{i2kx}. \quad (5.21)$$

It may be seen that the magnitude of the reflection coefficient in an ideal line is constant, but its phase varies with the position along the line, x. Frequently, it is convenient to refer the reflection coefficient to its value at $x = 0$:

$$\rho(\omega, x) = \rho(\omega, 0) e^{i2kx}. \quad (5.22)$$

From Eqs. 5.17, 5.21 and 5.22 it follows that the voltage at any point of the ideal line may be obtained from the complex amplitude of the incident and reflected voltage waves, U_{inc} and U_{ref}, and the reflection coefficient measured at the reference plane ($x = 0$), $\rho(\omega, 0)$.

The equivalent impedance $Z(\omega, x)$ is defined as the ratio between the voltage between the line conductors and the current flowing through them:

$$Z(\omega, x) = \frac{V(\omega, x)}{I(\omega, x)}. \quad (5.23)$$

Therefore, from Eqs. 5.13 and 5.14 it follows that:

$$Z(\omega, x) = \frac{V_{\text{inc}}(\omega, x) + V_{\text{ref}}(\omega, x)}{I_{\text{inc}}(\omega, x) - I_{\text{ref}}(\omega, x)}. \quad (5.24)$$

Replacing Eqs. 5.15, 5.16 and 5.20 in Eq. 5.24 it is easy to see that:

$$\frac{Z(\omega, x)}{Z_0} = \frac{1 + \rho(\omega, x)}{1 - \rho(\omega, x)}. \quad (5.25)$$

From the above, if an impedance $Z(\omega, x)$ is placed the plane x of an ideal line of characteristic impedance Z_0 the reflection coefficient is given by

$$\rho(\omega, x) = \frac{Z(\omega, x) - Z_0}{Z(\omega, x) + Z_0}. \quad (5.26)$$

Finally, if the reflection coefficient at x is written in terms of its value at the reference plane $x = 0$ it follows that:

$$\frac{Z(\omega, x)}{Z_0} = \frac{1 + \rho(0, x)e^{i2kx}}{1 - \rho(0, x)e^{i2kx}}. \tag{5.27}$$

Therefore, if the sample cell of impedance $Z_s(\omega)$ is connected at the end of the line, at the plane located at $x = L$, from Eq. 5.27 it follows that its value may be obtained from the reflection coefficient measured at the reference plane $x = 0$:

$$\frac{Z_s(\omega)}{Z_0} = \frac{1 + \rho(0, x)e^{i2kL}}{1 - \rho(0, x)e^{i2kL}}. \tag{5.28}$$

As explained in previous sections, for dielectric measurements it is often more convenient to determine the admittance $Y_s(\omega) = 1/Z_s(\omega)$:

$$Y_s(\omega) = \frac{1}{Z_0} \frac{1 - \rho(0, x)e^{i2kL}}{1 + \rho(0, x)e^{i2kL}}. \tag{5.29}$$

5.3.4 Transmission Lines under Pulsed Excitation

In dielectric measuring systems based on time-domain reflectometry (TDR), the excitation waveform is a pulse or a "step" signal. The information provided by these non-harmonic excitations is equivalent to that obtained from harmonic signals, through the application of Fourier and Laplace transforms.

The Laplace transforms of $V_{inc}(x, t)$ and $V_{ref}(x, t)$ are:

$$V_{inc}(s, x) = \int_0^{+\infty} V_{inc}(x, \tau)e^{-s\tau}d\tau \tag{5.30}$$

$$V_{ref}(s, x) = \int_0^{+\infty} V_{ref}(x, \tau)e^{-s\tau}d\tau \tag{5.31}$$

where the transform variable $s = \alpha + i\omega$ is termed the complex angular frequency. As usual, it is assumed that the excitation is zero for $t < 0$ and that the system is linear and causal.

Under these assumptions, which are always valid in DS measurement systems, $V_{inc}(\omega, x)$ and $V_{ref}(\omega, x)$ are obtained from the generalized Fourier transforms, that is, taking the limit of the Laplace transforms (5.30) and (5.31) when α, the real part of the transform variable s, tends to zero:

$$V_{inc}(\omega, x) = \lim_{\alpha \to 0} V_{inc}(s, x) \tag{5.32}$$

$$V_{ref}(\omega, x) = \lim_{\alpha \to 0} V_{ref}(s, x). \tag{5.33}$$

It must be remarked that the Laplace transform exists for a very wide class of functions. In consequence, the generalized Fourier transform may then be defined under much weaker conditions than the usual Fourier transform (in particular, it is not limited to square-integrable functions). This would not have been always possible if the usual Fourier transform were directly applied to $V_{inc}(x, t)$ and $V_{ref}(x, t)$, for instance, when the excitation signal is a step function.

In summary, once $V_{inc}(\omega, x)$ and $V_{ref}(\omega, x)$ are obtained from the generalized Fourier transform of the measured signals $V_{inc}(x, t)$ and $V_{ref}(x, t)$, it is possible to analyze the time-domain data from a system excited by non-harmonic waveforms ("pulse", "step", etc.) using the formalism developed for time-harmonic excitation in the previous sections.

5.4 Experimental Techniques

5.4.1 Temperature Control

The electric parameters of liquid samples are highly dependent on temperature; furthermore, the study of the dependence of the liquid properties with temperature provides valuable information. In consequence, temperature must be homogeneous in the entire sample volume, and must be measured to 0.1 °C and kept stable within 0.03 °C or better during the measurements corresponding to each isothermal dielectric spectrum.

For the samples studied in this book, the range of temperatures of interest is form room temperature to about 75 °C; therefore, a good quality immersion thermostat using water as thermostatic fluid will be found quite satisfactory. Depending on the sample cell design details, it may be connected to the thermostat using short lengths of silicone rubber tubing. Of course, an adequate rate of fluid circulation is necessary to ensure the temperature homogeneity and stability of the sample cell.

5.4.2 Sample Cells

Sample cells for low-frequency work usually consist on plane parallel electrodes supported in glass or PTFE (Teflon®). Platinized platinum ("platinum black") electrodes of the usual kind used for electrochemical work have been found by the authors to be quite satisfactory for the measurements described in this book. Several designs have been described in the scientific literature [5, 6].

The capacitance of the empty cell is usually of the order of a few picofarads to a few tens of picofarads. The volume of the sample must be adequate to ensure that the electrodes are fully submerged. Typical volumes are of the order of 25 ml.

The calibration of the cell is done with standard reference liquids; high purity cyclohexane (HPLC grade) is often used, since it has negligible conductivity and its permittivity is constant up to frequencies of several GHz ($\varepsilon' = 2.015$ at 298 K).

For work at higher frequencies, coaxial geometries are often used. The design must follow good design practices, in order to minimize losses and avoid any parasitic effects. The materials in contact with the sample must be also carefully chosen. The authors have obtained good results with stainless steel (type 316) and PTFE (Teflon®).

In all cases the design of the sample cell must allow an adequate rate of flow of the thermostatic fluid (usually of the order of a few liters/minute), and must minimize the thermal coupling of the sample with the surroundings. Double-wall designs are typical.

Several manufacturers provide high-frequency sample cells for liquids, and several designs have been described in the literature [7–9].

5.4.3 Sample Handling

Careful laboratory procedures are very important to obtain repeatable results. Samples must be handled with care, in order to avoid contamination and all glass elements must be clean. Proper cleaning of the sample cell is of paramount importance. An adequate cleaning procedure typically consists in the steps indicated in Box 5.1. All cleaning fluids must be of high purity (*pro analysis* grade or better).

Box 5.1: Cleaning Procedure for Sample Cells

1. Remove the sample.
2. Carefully rinse with ethyl alcohol (ethanol) to remove all the visible remnants of the sample.
3. Immerse in isopropyl alcohol for 5 min.
4. Empty the cell.
5. Immerse again in isopropyl alcohol for 10 min.
6. Empty the cell and keep in air until all traces of isopropyl alcohol have evaporated (typically not <30 min)!
7. Re-check visually!!!
8. Check the electrical parameters of the empty cell.

In all cases, the volume of the sample must be carefully controlled when the sample cell is filled, to insure that it is the same as the sample of the reference liquid used for calibration. In this way, any systematic errors associated to edge effects in the liquid surface are minimized.

Safety rules must be strictly followed. Since the samples may contain methanol, hydrocarbons and other dangerous substances, gloves and other personal protection elements must be always used. Moreover, adequate ventilation of the working area must be ensured, and means for the cleaning of spills must be available.

After the cleaning procedure is completed, the capacitance of the empty cell must be checked, to verify that its value coincides (within the stated uncertainty) with the calibration value. The measured value of the imaginary part of the capacitance must always be very low (less or equal to the value measured with the cyclohexane calibration sample) if the cleaning procedure was successful and all traces of isopropyl alcohol have evaporated.

5.4.4 Low-Frequency Measuring Instruments

Low-frequency dielectric measurements require the measurement of the admittance of the sample cell at a number of frequencies. Typical frequency ranges are from 20 Hz to 20 MHz. Typically, the system must be able to measure the magnitude of capacitances in the range from 20 to 200 pF with an uncertainty lower than 1%, with dissipation factors lower than 0.005.

Electronic instrumentation for this purpose integrates a dielectric measuring interface, a signal source and an amplitude/phase measuring system. This may be implemented with separate modules [10] or in a single instrument, such as an RCL meter, automated impedance bridge or impedance analyzer [11–13]. Of course, computer control and data acquisition is very useful, although not mandatory.

For the measurement of conductivity only, portable instruments (conductimeters) are very convenient, particularly at production plants. Some models include temperature measurement in the probe and a computer interface to download the data for further processing.

Specialized software may be used to extract the electrical parameters from the experimental data, although if necessary the user may write the necessary programs using spreadsheet software.

Needless to say, proper care, calibration and use of the instruments, according to the manufacturer's recommendations, are essential for accurate results. Particular attention should be paid to the temperature and humidity limits for storage and operation of the equipment, as well as the characteristics of the power mains supply (line voltage and frequency, etc.).

All connection cables and connectors must be of the highest quality, and proper periodic inspection and cleaning of the connectors (at least once a month) is a prerequisite for good results.

5.4.5 High-Frequency Measuring Instruments

Measurements above 20 MHz involve the generation of the excitation signal, and the measurement of the reflection coefficient. Typical uncertainties in the measurement of the reflection coefficient are less than ± 0.005 in magnitude, and ± 0.01 rad in phase (in both cases, after proper calibration). The instruments used with time-harmonic excitation signals are coaxial reflectometers, RF vector volt-meters and network analyzers [14]. For the separation of incident and reflected waves, high performance directional couplers are used (directivity >40 dB) [15]. An alternative, TDR, uses non-harmonic excitation sources (typically pulse or step signals with rise-time <10 ps), together with high speed digitizers (5 Gs/s and higher) [16]. Both kinds of systems are always computer controlled.

In all cases, specific calibration procedures must be carefully executed according to the manufacturer instructions. The electrical parameters of the sample are obtained from the experimental data using specialized software [17, 18], including parameters obtained from the calibration procedure. This usually involves the use of specific verification kits, as recommended by the manufacturer.

As indicated in the previous section, proper care, calibration and use of the instruments, according to the manufacturer recommendations, is essential for accurate results. All high-frequency connection cables must be of the highest quality and the connectors must be Type N (metrology-grade) or APC-7 [15]. Proper periodic inspection and cleaning of the connectors (at least once a month) is a prerequisite for good results. This includes the verification of the mechanical tolerances of the connectors using appropriate gages, according to manufacturer specifications.

5.4.6 Optical Measurements: Refractive Index

Measurements of refractive index in the visible range are usually made with refractometers based on the limiting angle principle, at the sodium D line (589 nm). The Abbe design is frequently used, and low-cost laboratory models can easily provide results with four-figure accuracy [19]. Of course, to achieve this level of accuracy the temperature of the sample must be kept constant to within $\pm 0.1\,^{\circ}\text{C}$.

5.5 Concluding Remarks

In order to achieve useful results with DS techniques, several important require-ments must be met in the design of the measuring system: adequate electronic instruments, proper design of the sample cells and good sample thermostatization.

Careful inspection and proper calibration at periodic intervals are necessary to ensure long term accuracy of the dielectric measuring system.

Laboratory procedures that are of critical importance in DS include the cali-bration and cleaning of sample cells, and proper sample handling.

Since some substances that may be found in samples from the different stages of biodiesel production are dangerous (i.e., methanol, strong bases, etc.), safety rules must be strictly followed, including the use of latex gloves and other personal protection elements.

References

1. Field RF (1966) Dielectric measurement techniques. In: von Hippel AR (ed) Dielectric materials and applications. MIT Press, Cambridge
2. Kremer F, Arndt M (1997) Broadband dielectric measurement techniques. In: Runt JP, Fitzgerald JJ (eds) Dielectric spectroscopy of polymeric materials. American Chemical Society, Washington DC
3. Lambri OA, Matteo CL, Mocellini RR, Sorichetti PA, Zelada GI (2008) Propiedades Viscoelásticas y Eléctricas de Sólidos y Líquidos. UNR Editora, Rosario
4. van Roggen A (1990) An overview of dielectric measurements. IEEE Trans Electr Insul 25:95–106
5. Kamimoto A (1972) A method for precise measurement of the dielectric constant of liquids in a wide frequency range. Rev Sci Instrum 43:763–765
6. Chute FS, Cervenan MR, Vermeulen FE (1978) Simple cell for the measurement of the radio frequency electrical properties of earth materials. Rev Sci Instrum 49:1675–1679
7. Schenkel CD, Sorichetti PA, Romano SD (2005) Electrodos intercambiables para medir propiedades eléctricas en líquidos. An Asoc Fis Argent 17:283–287
8. Damaskos Inc (2009) Cavities and resonators. http://www.damaskosinc.com/cavitiy.htm. Accessed 11 Nov 2009
9. Agilent Technologies (2008) 85070E dielectric probe kit 200 MHz to 50 GHz Technical Overview. http://cp.literature.agilent.com/litweb/pdf/5989-0222EN.pdf. Accessed 11 Nov 2009
10. Sorichetti PA, Matteo CL (2007) Low-frequency dielectric measurements of complex fluids using high-frequency coaxial sample cells. Measurement 40:437–449
11. Pratt GJ, Smith MJA (1982) A wide-band bridge for the measurement of the complex permittivity of polymeric solids and other materials. J Phys E Sci Instrum 15:927–933
12. Bennani H, Pilet JC (1995) A simple apparatus to plot the variations of the dielectric permittivity versus temperature in the range of frequencies 10 Hz–100 kHz. IEEE Trans Instrum Meas 41:438–440
13. Agilent Technologies (2008) Solutions for measuring permittivity and permeability with LCR meters and impedance analyzers. Application Note 1369-1. http://cp.literature.agilent.com/litweb/pdf/5980-2862EN.pdf. Accessed 11 Nov 2009
14. Agilent Technologies (2009) Network analyzer selection guide. http://cp.literature.agilent.com/litweb/pdf/5989-7603EN.pdf. Accessed 11 Nov 2009
15. Agilent Technologies (2007) Agilent RF and microwave test accessories Catalog 2006/07. http://cp.literature.agilent.com/litweb/pdf/5968-4314EN.pdf. Accessed 11 Nov 2009
16. Material Sensing & Instrumentation, Inc. (2007) Broadband dielectric spectroscopy. http://www.msi-sensing.com/broadband_dielectrics.htm Accessed 11 Nov 2009
17. Damaskos Inc (2009) Software packages http://www.damaskosinc.com/software.htm. Accessed 11 Nov 2009
18. Agilent Technologies (2008) 85071E Materials measurement software Technical Overview. http://cp.literature.agilent.com/litweb/pdf/5988-9472EN.pdf. Accessed 11 Nov 2009
19. Atago Abbe Refractometer model NAR-1T User Manual (2010) http://www.atago.net. Accessed 11 Nov 2009

Chapter 6
Dielectric Techniques
for the Characterization of Raw Materials
and Effluents in Biodiesel Production

6.1 Introduction

In this chapter, the application of dielectric measurements to pure and used vegetable oils of different origins will be discussed. The conductivity and low-frequency permittivity values as a function of temperature are used to characterize the oils used as raw materials for biodiesel production, including the pre-treatment (when necessary).

In addition, the relation between the electrical properties of the effluents and the degree of advance of the washing process are discussed in connection with the production process of biodiesel.

6.2 Dielectric Properties of Vegetable Oils

As explained in Chap. 2, vegetable oils are the basic raw material for biodiesel production. In this section the dielectric properties of oils from different origins will be presented:

- Rapeseed
- Sunflower
- Soybean
- Corn
- Olive

Three typical broadband isothermal dielectric spectra (real part) are shown in Fig. 6.1. The data correspond to soybean oil at temperatures of: 25, 55 and 75°C (298.15, 328.15 and 348.15 K). It may be seen that ε' is essentially constant in the entire frequency range (six decades) up to 10 MHz, but clearly depends on temperature. The data shown are unfiltered and correspond to an actual measurement

S. D. Romano and P. A. Sorichetti, *Dielectric Spectroscopy in Biodiesel Production and Characterization*, Green Energy and Technology, DOI: 10.1007/978-1-84996-519-4_6,
© Springer-Verlag London Limited 2011

Fig. 6.1 Dielectric spectra
(real part) of soybean oil at
three different temperatures.
Notice the downward noise
spike at 50 Hz

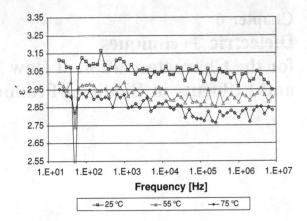

run; for instance, the effects of power line noise at the frequency of 50 Hz are
clearly visible as a downward spike in the three spectra.

At each absolute temperature the dielectric parameters are obtained from the
fitting to a dielectric model with $\varepsilon'(T)$ independent of ω and negligible $\varepsilon_{pol}{}''$, plus a
conductivity term $\sigma(T)/\omega$ in the imaginary part of the complex permittivity.

$$\varepsilon_r(\omega, T) = \varepsilon'(T) - i\frac{\sigma(T)}{\varepsilon_0\omega}. \tag{6.1}$$

Since thermal agitation tends to oppose the polarization associated to molecular
orientation, ε', may be expected to decrease with temperature. Therefore, for
temperatures in the range from 295 to 350 K it will be modeled by a linear
function of the absolute temperature:

$$\varepsilon'(T) = a - bT. \tag{6.2}$$

Moreover, the conductivity σ is negligible below 25°C but increases very
rapidly with temperature. In the range of absolute temperatures presented in this
chapter (295–350 K), the dependence may be modeled by an exponential function:

$$\sigma(T) = ce^{dT}. \tag{6.3}$$

The result of the fitting of Eq. 6.1 to experimental data at 25°C is shown in
Figs. 6.2 (real part) and 6.3 (imaginary part). The square symbols correspond to
individual data points, the continuous lines to the fitted values, and the two dashed
lines are the limits of the root mean square (RMS) error bands. The fitting
parameters at 25°C: $\varepsilon' = 3.03 \pm 0.01$; $\varepsilon'' = 0.01 \pm 0.02$. It is clear that at this
temperature the dielectric losses are negligible. On the contrary, experimental
results for the imaginary part of the complex permittivity at 75°C (Fig. 6.4) may
be fitted assuming a value of $\sigma = 8.3 \times 10^{-11}$ S/m.

The fitted values of permittivity and conductivity ε' and σ for several vegetable
oils at 25, 55 and 75°C are plotted in Figs. 6.5–6.10.

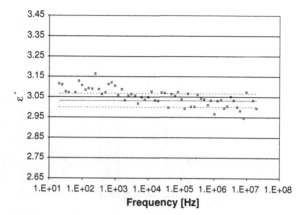

Fig. 6.2 Dielectric spectra (real part) of soybean oil at 298.15 K (25°C)

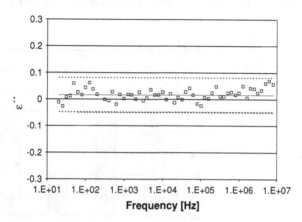

Fig. 6.3 Dielectric spectra (imaginary part) of soybean oil at 298.15 K (25°C)

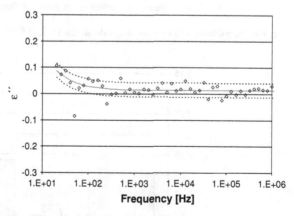

Fig. 6.4 Dielectric spectra (imaginary part) of soybean oil at 348.15 K (75°C)

It must be remarked that the conductivities of both pure soybean and corn oils at 25°C are below the minimum of the measuring system, 10^{-11} S/m (cf. Fig. 6.8).

The parameters of the dependence on temperature of ε' (a and b) and σ (c and d) (cf. Eqs. 6.2 and 6.3) are summarized in Tables 6.1 and 6.2, together with their

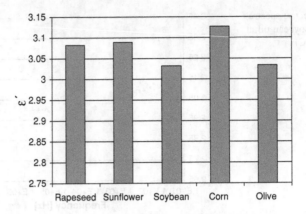

Fig. 6.5 Permittivity (ε') of vegetable oils at 298.15 K (25°C)

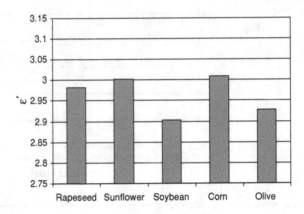

Fig. 6.6 Permittivity (ε') of vegetable oils at 328.15 K (55°C)

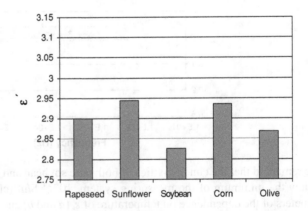

Fig. 6.7 Permittivity (ε') of vegetable oils at 348.15 K (75°C)

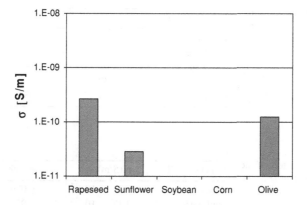

Fig. 6.8   Conductivity (σ) of vegetable oils at 298.15 K (25°C)

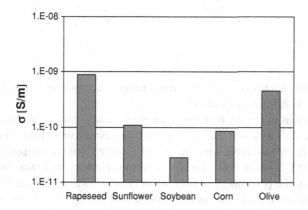

Fig. 6.9   Conductivity (σ) of vegetable oils at 328.15 K (55°C)

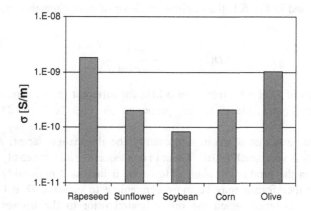

Fig. 6.10   Conductivity (σ) of vegetable oils at 348.15 K (75°C)

Table 6.1 Parameters of the linear fit of $\varepsilon'(T)$ in the range from 295 to 350 K

Vegetable oil	a	b (1/K)	R^2
Rapeseed	4.16	0.0036	0.994
Sunflower	3.94	0.0029	0.994
Soybean	4.33	0.0043	0.994
Corn	4.30	0.0039	0.999
Olive	4.04	0.0034	0.997

Table 6.2 Parameters of the exponential fit of $\sigma(T)$ in the range from 295 to 350 K

Vegetable oil	c (S/m)	d (1/K)	R^2
Rapeseed	2.43×10^{-15}	0.039	0.998
Sunflower	3.17×10^{-16}	0.039	0.984
Soybean	7.55×10^{-19}	0.053	0.984
Corn	1.25×10^{-17}	0.048	0.990
Olive	3.25×10^{-16}	0.043	0.998

correlation coefficients (R^2). The estimated relative uncertainty in the parameters a and b are of 1 and 3%, respectively.

It is important to remark that the parameters a and c correspond numerically to the values of $\varepsilon'(T)$ and $\sigma(T)$ extrapolated to 0 K, respectively. However, dielectric measurements in vegetable oils are usually carried out at temperatures above 283 K ($\approx 10°C$) since they usually become cloudy at lower temperatures (depending on the percentage of fatty acids present). Therefore, the values of the dielectric parameters of the model represented by Eq. 6.1 have no physical meaning if extrapolated to lower temperatures.

The dissipation factor D, defined in Eq. 4.32, is also used to compare dielectric losses, particularly for technological applications. According to the dielectric model presented in Eq. 6.1, the dissipation factor of pure vegetable oils will vary as ω^{-1}:

$$D(\omega, T) = \frac{\sigma(T)}{\omega \, \varepsilon_0 \, \varepsilon'(\omega, T)}. \tag{6.4}$$

The values of D at a frequency of 5 kHz for different (pure) vegetable oils at 298.15, 328.15 and 348.15 K are presented in Figs. 6.11, 6.12 and 6.13, respectively.

It may be seen that at room temperature the dissipation factor, D, of pure vegetable oils is very small ($<10^{-3}$), and in consequence, it is not easily measured, particularly in the production plants. However, it increases noticeably with temperature and therefore it may be more convenient to measure D at higher temperatures, for instance, when the oil is heated prior to the transesterification reaction.

Fig. 6.11 Dissipation factor
(*D*) at 5 kHz of pure vegetable oils at 298.15 K (25°C)

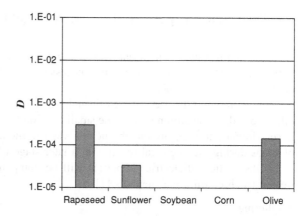

Fig. 6.12 Dissipation factor
(*D*) at 5 kHz of pure vegetable oils at 328.15 K (55°C)

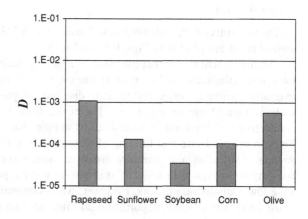

Fig. 6.13 Dissipation factor
(*D*) at 5 kHz of pure vegetable oils at 358.15 K (75°C)

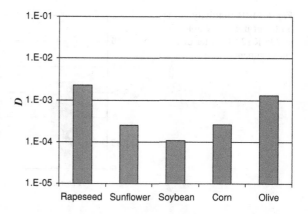

6.3 Dielectric Properties of Treated Vegetable Oils

As explained in Chap. 2, the treatment of vegetable oils (new or used) prior to the transesterification reaction is sometimes necessary in order to achieve a reasonable quality of the end product.

The treatment of the used frying oils includes a filtration step to homogenize the oil, followed by the elimination of the small fraction of water it usually contains. It is very important to ensure the absence of water in the oil to reduce the formation of soaps during the chemical reaction at a later stage of the process.

In this section, dielectric properties will be compared for used vegetable oil treated by three different methods [1]:

- Filtering
- Drying (at 80°C)
- Distillation in a partial vacuum at room temperature: final pressure below 10^2Pa (≈ 1 mmHg).

The permittivity, ε', and dissipation factor, D, at 5 kHz at 295 K (22°C) of used vegetable oil are plotted in Figs. 6.14 and 6.15.

The reader will notice that the value of ε' is very nearly the same for the all the treatments (the scale in Fig. 6.14 is linear). On the contrary, D is an order of magnitude lower for dried and vacuum distilled samples than for filtered oil (the scale is logarithmic in Fig. 6.15). Therefore, from Eq. 6.4 it follows that the conductivity of dried and vacuum distilled samples has been significantly reduced. These results indicate that filtering eliminates only solid particles but drying and vacuum distillation also removes dissolved conductive contaminants. From the comparison of Figs. 6.11 and 6.15 it is easy to see that pure, unused vegetable oils have lower dissipation factors. However, when comparing the dissipation factors of used and pure oils, it is important to take into account that the former are usually mixtures of oils from several origins.

Fig. 6.14 Permittivity (ε') at 5 kHz of used vegetable oils at 295 K (22°C) after different treatments

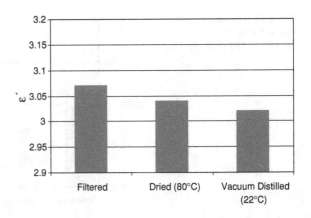

Fig. 6.15 Dissipation factor (*D*) at 5 kHz of used vegetable oils at 295 K (22°C) after different treatments

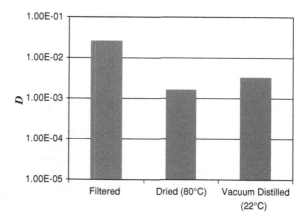

Fig. 6.16 Permittivity (ε') of the effluents of each washing step

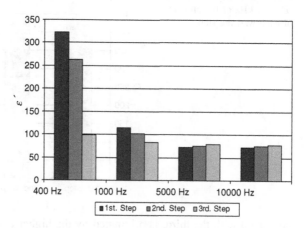

6.4 Dielectric Properties of Effluents from Biodiesel Production

The main effluent form biodiesel production is the water from the washing steps, as explained in Chap. 2. The measurements of the dissipation factor and conductivity of the effluents gives a clear indication of the presence of polar molecules such as methanol and also ionic species such as hydroxides used as catalysts. Also, acids added in the first washing step contribute to the dielectric loss of the effluent.

The permittivity, ε', of the effluents of each washing step is plotted at several frequencies in Fig. 6.16. It may seen that the measured value of the permittivity at the lower frequencies (1 kHz and below) is noticeably higher than the permittivity of water (≈ 80) due to the effects of interfacial polarization on the electrodes of the sample cell (cf. Sect. 4.6) At higher frequencies (from 5 kHz) ε' remains practically constant at the value of the permittivity of water. The large difference in the measured values of ε' at low frequencies, for the first and second washing steps, as

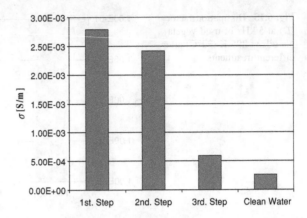

Fig. 6.17 Conductivity (σ) of the effluents of each washing step

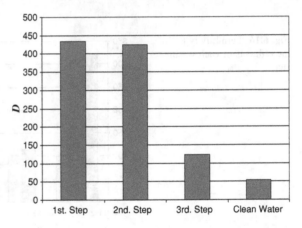

Fig. 6.18 Dissipation factor (D) at 1 kHz of the effluents of each washing step

compared with the third, is originated by the higher concentration of ions (giving rise to a stronger interfacial polarization effect).

The conductivity of the effluents (σ) is plotted in Fig. 6.17. It may be seen that its value decreases steeply with each washing step, down to a value close to that of the water used for the washing. It must be remembered that the conductivity of the clean water is lower limit of the conductivity of the effluents. The measured value of σ is practically independent of the frequency, and depends clearly on the concentration of ions in the effluents. It may be easily seen in the figure that there is a sharp drop in the value of conductivity between the second and third washing steps, corresponding to the decrease of electrode polarization effects that were mentioned in connection to Fig. 6.16.

The dissipation factor, D, measured at 1 and 10 kHz, is plotted in Figs. 6.18 and 6.19. At 1 kHz the measured value of D shows the effects of interfacial polarization, as follows from Eq. 6.4, since the conductivity σ is constant, but the measured value of ε' increases at lower frequencies. In consequence, the behavior of D at low frequencies departs form the ω^{-1} dependence that is usual when there are no polarization effects [2, 3].

Fig. 6.19 Dissipation factor
(D) at 10 kHz of the effluents
of each washing step

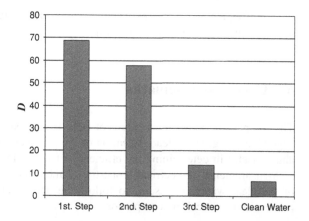

The progressive reduction of the conductivity and interfacial polarization in the effluents give a clear indication of the removal of the methanol and the hydroxide (catalyst) from the mixture of methyl esters (FAME). When the measured values in the effluents are close to those of the clean water, it may be assumed that the washing is adequate, as confirmed by dielectric measurements on FAME [1].

It must be remarked that the results presented in this section were obtained from measurements on samples kept at room temperature, as is the usual practice in industrial settings. Results may differ significantly if samples are measured at much higher (or lower) temperatures.

6.5   Refractive Index of Vegetable Oils

As explained in Sect. 4.6.1, in the optical range the permittivity is controlled by electronic polarization processes. In this range, measurements of refractive index are used to determine the permittivity (cf. Sect. 4.3). Since in pure vegetable oils the attenuation at visible wavelengths is small the low-loss approximation, Eq. 4.44, is valid and ε' is given by the square of the refractive index, n. Values shown in Table 6.3 were measured at the wavelength of the D line of sodium (589 nm) and correspond to a value of permittivity around 2.18. The difference in

Table 6.3 Refractive index of pure vegetable oils

Vegetable oil	n_D (20°C)	n_D (40°C)
Rapeseed	1.477	1.470
Sunflower	1.479	1.472
Soybean	1.478	1.471

It is worth mentioning that the refractive index of used vegetable oils at room temperature (295 K) is slightly lower ($n_D = 1.47$) [1]

the low-frequency values (between 3 and 3.1) is originated by the relaxation processes in the infrared band (cf. Eq. 4.39).

6.6 Concluding Remarks

The comparison between the results of dielectric measurements of pure and treated vegetable oils gives a clear indication of the presence of water and contaminants, and is useful in determining the efficiency of the different treatment procedures.

Measurements of electrical properties of the effluents, as compared with those of the clear water, are useful to follow the degree of advance of the washing process and to check its completion.

References

1. Sorichetti PA, Romano SD (2005) Physico-chemical and electrical properties for the production and characterization of biodiesel. Phys Chem Liq 43(1):37–48
2. Piper JD, von Hippel AR (1954) Liquids dielectrics. In: von Hippel AR (ed) Dielectric materials and their applications, 1st edn. Wiley, New York
3. Coelho R, Aladenize B (1993) Polarization, permittivité et relaxation. In: Les Dielectriques, 1st edn. Hermes, Paris

Chapter 7
Application of Dielectric Spectroscopy to the Characterization of FAME in Biodiesel Production

7.1 Introduction

As explained in Chap. 2, during the production process, the mixture of fatty acid methyl esters (FAME) must be separated from glycerin and then purified in several steps.

In this chapter, the electrical properties (permittivity and conductivity) at different temperatures of FAME from different production stages will be presented. The results from these measurements give a quantitative indication of the advance of the purification process and the conversion to biodiesel.

7.2 Dielectric Model of FAME and Biodiesel

Due to the molecular structure of biodiesel (nonpolar molecules), there are no relaxation processes up to frequencies of several GHz [1]. In consequence, a dielectric model akin to vegetable oils will be used [cf. Eq. 6.1] where the permittivity is constant with frequency, but decreases linearly with temperature [cf. Eq. 6.2]. Conductivity of biodiesel is found to increase noticeably at higher temperatures, and experimental data may be fitted to an Arrhenius law:

$$\sigma(T) = \sigma_0 e^{-\frac{d}{T}} \tag{7.1}$$

whereas the fitting parameter d may be considered as proportional to an activation energy ΔE:

$$d = \frac{\Delta E}{k_B} \tag{7.2}$$

S. D. Romano and P. A. Sorichetti, *Dielectric Spectroscopy in Biodiesel Production and Characterization*, Green Energy and Technology, DOI: 10.1007/978-1-84996-519-4_7, © Springer-Verlag London Limited 2011

(k_B is Boltzmann constant). This seems to indicate that conductivity in FAME free of contaminants (biodiesel) is due to a thermally activated process with an activation energy of approximately 0.21 eV. This is a reasonable value for this kind of molecules [1].

The same dielectric model is applied to the mixture of fatty acid methyl esters (FAME) during the washing process. It is important to note, however, that conductivity is noticeably higher in FAME prior to the first washing step (i.e., immediately after the separation of glycerin), due to the presence of ionic molecules (from the catalyst) and methanol. Therefore, the dependence of conductivity on temperature for non-washed FAME departs from the Arrhenius behavior described by Eq. 7.1, as it will be discussed in Sect. 7.3.2.

7.3 Dielectric Properties of FAME

7.3.1 Permittivity

The initial (non-washed) mixture of fatty acid methyl esters (FAME) contains methanol and catalyst remnants. It is important to remark that the dependence on temperature of the permittivity of polar molecules (such as methanol) is stronger than for nonpolar substances. Therefore, at any given temperature, the successive washing steps will result in a reduction of the permittivity, as the concentration of polar impurities is reduced.

In Fig. 7.1, the permittivity of non-washed FAME is plotted as a function of temperature. This should be compared with Fig. 7.2 that shows the permittivity of FAME after the first and second washing steps. In both figures, the volumetric alcohol-to-oil ratio, R, is the value generally used in practice (0.25), resulting in a relatively low concentration of polar impurities in the non-washed FAME.

Fig. 7.1 Permittivity (ε') of non-washed FAME ($R = 0.25$)

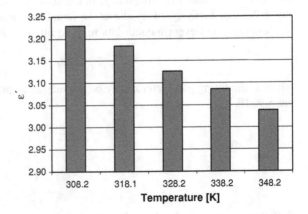

Fig. 7.2 Permittivity (ε') of
FAME after the first and
second washing steps
($R = 0.25$)

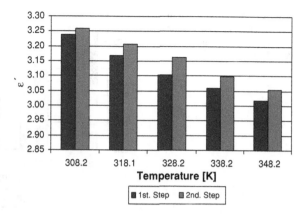

Fig. 7.3 Permittivity (ε') of
non-washed FAME
($R = 0.40$)

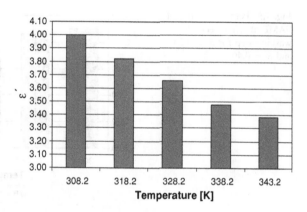

Therefore, the temperature dependence of the permittivity of FAME after the
successive washing steps (Fig. 7.2) is similar to non-washed FAME (Fig. 7.1).

The permittivity of non-washed FAME with high alcohol-to-oil ratio
($R = 0.40$) is shown in Fig. 7.3. As it is evident from the comparison of Figs. 7.1
and 7.3, when the volumetric alcohol-to-oil ratio, R, is higher than 0.25, the
permittivity of non-washed FAME is noticeably higher and decreases more steeply
with temperature, indicating that the permittivity is strongly influenced by polar
impurities (mainly methanol).

If the washing process is efficient, after the first washing step the difference
between the permittivity values for $R = 0.25$ and $R = 0.40$ at each temperature
will be smaller, as it follows from Fig. 7.4.

After the second washing step, the permittivity values will be practically the
same, independently of the alcohol-to-oil ratio, as shown in Fig. 7.5. These results
complement the measurements on the effluents of the washing process discussed in
Sect. 6.4.

Fig. 7.4 Permittivity (ε') of FAME after the first washing step for $R = 0.25$ and $R = 0.40$

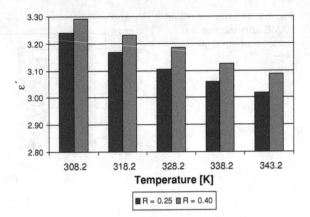

Fig. 7.5 Permittivity (ε') of FAME after the second washing step for $R = 0.25$ and $R = 0.40$

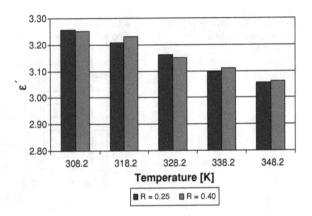

7.3.2 Conductivity

As indicated in the previous paragraphs, non-washed FAME contains methanol and catalyst remnants, and in consequence the conductivity will be higher. Moreover, it may be seen from Fig. 7.6 that the conductivity falls rapidly with temperature. This may be understood as a consequence of the reduction of the mobility with increasing temperature, of the mobile charge carriers (due to the impurities) present in non-washed FAME.

After the first washing step, most of the impurities are removed, resulting in a significant reduction in conductivity, as seen in Figs. 7.6 and 7.7. Moreover, the conductivity after the first washing step is found to increase exponentially with temperature. This implies that the conductivity no longer depends on the mobility of the mobile charge carriers present in non-washed FAME, but, similarly to biodiesel, it is controlled by a thermally activated mechanism, described by Eq. 7.1.

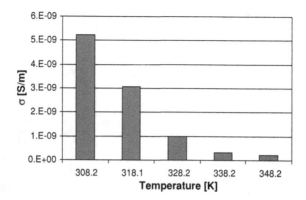

Fig. 7.6 Conductivity (σ) of non-washed FAME ($R = 0.25$)

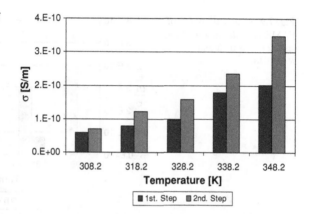

Fig. 7.7 Conductivity (σ) of FAME after the first and second washing steps ($R = 0.25$)

Since the first washing step is carried out with acidified water, to neutralize the remaining (basic) catalyst after this washing step, FAME (mainly a nonpolar substance) contains non-dissociated acid molecules. In consequence, the water added in the second washing step dissociates the acid in FAME, resulting in an increase in conductivity, as seen in Fig. 7.7.

The conductivity of non-washed FAME with high alcohol-to-oil ratio ($R = 0.40$) is shown in Fig. 7.8. Comparing Figs. 7.6 and 7.8, the conductivity of non-washed FAME for high alcohol-to-oil ratio ($R = 0.40$) is several orders of magnitude higher than for $R = 0.25$ and also shows a steady diminution with temperature. The higher value of conductivity is easy to understand taking into account that the additional mobile charge carriers originated from the significant excess in methanol for $R = 0.40$ in comparison with $R = 0.25$.

After the first washing step, the conductivity of FAME for high alcohol-to-oil ratio as a function of temperature presents a minimum, as it may be seen in Fig. 7.9. This indicates that, as distinct from FAME with $R = 0.25$, at high R values the first washing step does not eliminate all the contaminants. Indeed, at lower temperatures, conductivity is still controlled by the mobility of charge carriers (that decreases with temperature), and at higher temperatures the

Fig. 7.8 Conductivity (σ) of non-washed FAME ($R = 0.40$)

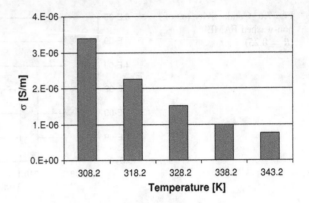

Fig. 7.9 Conductivity (σ) of FAME after the first washing step for $R = 0.25$ and $R = 0.40$

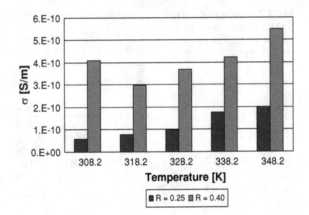

Fig. 7.10 Conductivity (σ) of FAME after the second washing step for $R = 0.25$ and $R = 0.40$

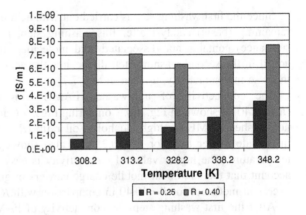

conductivity increases exponentially, dominated by a thermally activated process in FAME molecules as explained in previous paragraphs.

The dependence of conductivity with temperature for $R = 0.40$ follows the same general trend after the second washing step, as it may be seen in Fig. 7.10.

Fig. 7.11 Permittivity (ε') of biodiesel as function of temperature

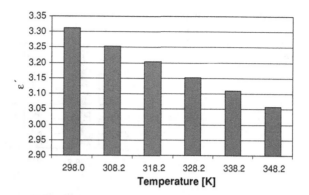

However, although conductivity values are significantly lower, mobility still controls the conductivity at lower temperatures.

7.4 Dielectric Properties of Biodiesel

7.4.1 Permittivity

The permittivity of biodiesel after the third washing step, when the contaminants have been removed from FAME, is shown in Fig. 7.11. The permittivity at 308 K (25°C) has the value $\varepsilon' = 3.32 \pm 0.05$. This value is characteristic of biodiesel, practically independent from the oils used as raw materials [2]. It may be seen that the permittivity fits very well to a linear function of temperature [cf. Eq. 6.2], similarly to pure vegetable oils, as indicated in Sect. 7.2. For the data presented in Fig. 7.11, the slope ($d\varepsilon'/dT$) is 0.0050 ± 0.0001, with a correlation coefficient (R^2) greater than 0.998. These results are summarized in Box 7.1.

Finally, it is worth mentioning that for the usual alcohol-to-oil ratio ($R = 0.25$), the permittivity of FAME (non-washed and also after the different washing steps) is practically the same as biodiesel, as it may be seen in Fig. 7.12.

Box 7.1 Permittivity of Biodiesel as a Function of Temperature

$$\varepsilon'(298K) = 3.32 + / - 0.05$$

$$d\varepsilon'/dT = 0.0050 + / - 0.0001$$

Fig. 7.12 Permittivity (ε') of non-washed FAME and after successive washing steps ($R = 0.25$)

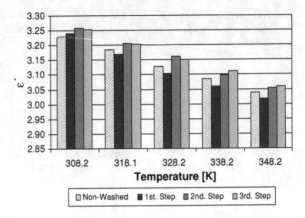

Fig. 7.13 Conductivity (σ) of biodiesel as function of temperature

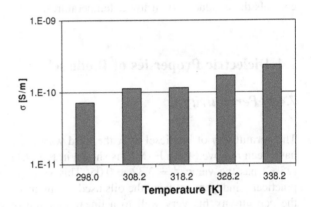

The permittivity of FAME after the third washing step is practically equal to the value for biodiesel, as explained above.

7.4.2 Conductivity

As indicated in Sect. 7.2, after removal of contaminants the conductivity of biodiesel is very low and controlled by a thermally activated process with activation energy of approximately 0.21 eV. The data shown in Fig. 7.13 (in a logarithmic scale) fit to the model in Eq. 7.2 with a correlation coefficient (R^2) >0.97. It is important to remark that measured conductivity will be higher in the presence of low concentrations of contaminants, as will be discussed in the next chapter.

Figure 7.14 shows (in a logarithmic scale) the conductivity values for non-washed FAME and after the successive washing steps, for the usual alcohol-to-oil ratio ($R = 0.25$). The figure summarizes the behavior of conductivity as a function

Fig. 7.14 Conductivity (σ) of non-washed FAME and after successive washing steps ($R = 0.25$)

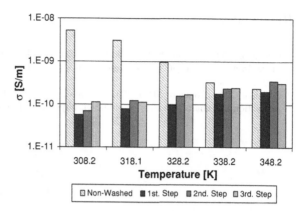

of temperature for non-washed FAME and FAME after the washing steps, including biodiesel.

7.5 Refractive Index

In the optical range, the permittivity is controlled by electronic polarization processes. As explained in Sect. 4.6.1, measurements of refractive index are used to determine the permittivity (cf. Sect. 4.3). The attenuation at visible wavelengths is small in biodiesel, and in consequence the low-loss approximation, Eq. 4.44, is valid and ε' is given by the square of the refractive index, n.

At room temperature (295 K), the refractive index of biodiesel, measured at the D line of sodium is $n_D = 1.4550 \pm 0.0005$ [2], which corresponds to a value of permittivity $\varepsilon_\infty = 2.117$. The difference with the low-frequency value $\varepsilon_0 = 3.32$ is originated from the relaxation processes in the infrared band [cf. Eq. 4.39].

It must be remarked that the refractive index of FAME remains unchanged after the successive washing steps, including biodiesel (within the stated uncertainty). This is reasonable since the refractive index of FAME depends essentially on the electronic polarization of the molecules of fatty acid esters. On the contrary, low-frequency dielectric measurements are also sensitive to polarization effects due to the presence of contaminants.

7.6 Concluding Remarks

Permittivity and conductivity measurements as a function of temperature are relevant for the production and characterization of biodiesel. In the first place, they provide relevant information on the efficiency of the successive washing steps of FAME. Furthermore, the characteristic electric parameters of biodiesel make it possible to detect contaminants in the final product, as discussed in the next chapter.

References

1. González Prieto LE, Sorichetti PA, Romano SD (2008) Electric properties of biodiesel in the range from 20 Hz to 20 MHz—comparison with fossil diesel fuel. Int J Hydrog Energy 33:3531–3537
2. Sorichetti PA, Romano SD (2005) Physico-chemical and electrical properties for the production and characterization of biodiesel. Phys Chem Liq 43(1):37–48

Chapter 8
Relations Between the Properties Required by International Standards and Dielectric Properties of Biodiesel

8.1 Introduction

As explained in Chap. 3, the measurement of a number of properties of biodiesel (BD) is required to verify the conformance of the final product to international standards. Among these parameters, this chapter will explore the relation between flash point (FP), methanol content (MC) and electrical properties.

Flash point is very important from the standpoint of safety in BD distribution and storage, since higher values lead to a safer handling of the fuel. The minimum value required by standards [1, 2] is usually 100°C or higher.

Methanol is one of the raw materials for the production of BD, as explained in Chap. 2. However, since it is impure, standards require the free methanol content in BD to be lower than 0.2%. Therefore, the excess of alcohol must be removed from the mix of methyl esters by an adequate purification process.

It is very important to take into account that free methanol content and flash point are closely related. Since the FP of pure methanol is 11°C, it is not surprising that FP of biodiesel will be lower than the minimum value indicated by the standards if MC exceeds the maximum allowed concentration of 0.2%.

Measurements of electrical properties (EP) carried out at different temperatures show a clear dependence between MC and the permittivity and conductivity of the samples. Although they are not required by international standards, they provide a convenient, low-cost alternative for the verification of the required parameters.

8.2 Determination of Flash Point

Flash point measurements are usually carried out using the Pensky-Martens Closed Cup Tester according to ASTM D 93 [3]. An outline of the Pensky-Martens

S. D. Romano and P. A. Sorichetti, *Dielectric Spectroscopy in Biodiesel Production and Characterization*, Green Energy and Technology, DOI: 10.1007/978-1-84996-519-4_8,
© Springer-Verlag London Limited 2011

Fig. 8.1 Outline of the Pensky-Martens closed cup test apparatus (side view)

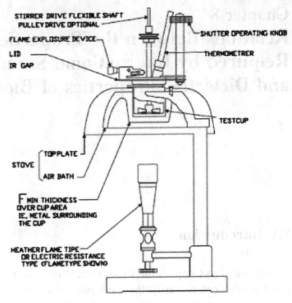

Fig. 8.2 Outline of the Pensky-Martens closed cup test apparatus (top view)

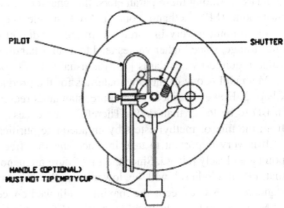

apparatus, in accordance with the design illustrated in ASTM D93, is shown in Figs. 8.1 and 8.2.

The test sample is placed in a brass test cup, of specified dimensions, that is fitted with a cover of specified dimensions. The cup is heated and the sample is stirred at a specified rate. At regular intervals, the stirring is interrupted and at the same time an ignition source is directed into the test cup, until a flash is detected.

The flash point is defined as the lowest temperature at which application of an ignition source causes the vapors of the test sample to ignite under the specified conditions indicated by the standard. For the purposes of FP determination, it is considered that the test sample has flashed when a flame appears and instantaneously propagates itself over the entire surface. The test conditions set by the standard [3] are indicated in Box 8.1.

Box 8.1 Test Conditions for FP Determination by the Pensky-Martens Method According to ASTM D 93

The temperature of the sample must increase at a rate of 5–6°C/min. The ignition source must be applied at each temperature increase of 2°C, beginning at a temperature of 23 ± 5°C below the expected flash point. The required sample volume is 70 ml.

8.3 Correlation Between Flash Point and Methanol Content in Biodiesel

Measurements show that there is a strong correlation between flash point and methanol content (Fig. 8.3). The following expression is valid for methanol content greater than or equal to 0.2%:

$$F = 38M^{-0.6}. \tag{8.1}$$

In (8.1), F is the flash point (in °C) and M is the methanol content (%V/V), and $M \geq 0.2$. The correlation coefficient (R^2) is greater than 0.99.

It is important to remark that the measured flash point value at the maximum methanol content set by international standards (0.2%) corresponds to the minimum allowable flash point (100°C) according to some international standards. Therefore, compliance with the flash point requirement is also a good indicator of compliance with maximum methanol content. As it is easy to see in Fig. 8.3, a small excess in methanol content originates a sharp decline in flash point of biodiesel.

To obtain the experimental values of FP as a function of MC presented in this chapter, biodiesel samples prepared by the process described in Chap. 2 were mixed with known volumes of methanol (analytical grade). The raw materials for the production of the biodiesel were soybean oil and methanol, using sodium

Fig. 8.3 Flash point as a function of methanol content in biodiesel

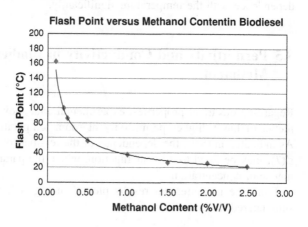

hydroxide as a catalyst, with the usual alcohol-to-oil ratio, $R = 0.25$. The first washing step was made with water acidified with hydrochloric acid. Finally, to check the quality of biofuel, the characterization properties required by international standards were measured.

8.4 Temperature Dependence of Electrical Properties of Biodiesel

In the frequency range of interest in this chapter, from about 20 Hz to 2 MHz, no dielectric relaxation processes are observed in biodiesel; therefore, ε' may be considered as independent of ω and ε_{pol}'' is negligible. On the other hand, the presence of free charge carriers due to the presence of contaminants and thermal effects in BD originate dissipative effects and the conductivity term σ/ω must be included in the imaginary part of the complex permittivity. Moreover, since thermal agitation tends to oppose the polarization associated to molecular orientation, ε', may be expected to decrease with temperature.

In summary, in this chapter the dependence of electrical properties of BD with the absolute temperature T and the angular frequency of the excitation, ω, will be fitted to the following expression:

$$\varepsilon(\omega, T) = \varepsilon'(T) - i\frac{\sigma(T)}{\varepsilon_0 \omega}. \tag{8.2}$$

The real part $\varepsilon'(T)$ will be assumed to have a linear dependence on temperature:

$$\varepsilon'(T) = a - bT \tag{8.3}$$

where the parameters a and b depend on the BD composition and also are affected by the presence of contaminants. On the contrary, the conductivity of pure BD is very low, but increases exponentially with temperature [4]; moreover, the presence of contaminants changes the room temperature value of conductivity and its dependence with the temperature significantly.

8.5 Permittivity and Conductivity in Biodiesel Containing Methanol

Regarding electrical properties, experimental data show that biodiesel containing methanol has a higher permittivity at room temperature than pure BD [4, 5]. As indicated in (8.3), the dependence of the real part of the complex permittivity, $\varepsilon'(T)$, may be fitted to a linear function, where the parameters a and b depend on methanol concentration.

From the measurement results plotted in Fig. 8.4, the fitting parameters are summarized in Table 8.1.

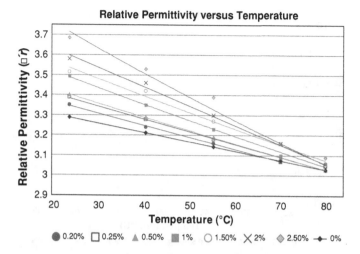

Fig. 8.4 Relativity permittivity versus temperature as a function of methanol content in biodiesel

Table 8.1 Parameters of the linear fit and correlation coefficient for permittivity versus temperature at different values of MC in biodiesel	Methanol concentration in biodiesel (%V/V)	a	b	R^2
	0.00	3.398	0.0046	0.999
	0.20	3.490	0.0059	0.998
	0.25	3.534	0.0062	0.988
	0.50	3.544	0.0061	0.998
	1.00	3.645	0.0074	0.994
	1.50	3.747	0.0084	0.996
	2.00	3.828	0.0095	0.993
	2.50	3.997	0.0117	0.949

From Fig. 8.4 it may be seen that the rate of decrease of permittivity with temperature increases with methanol content. Also, the fitting of permittivity data as a linear function of temperature is better at lower methanol concentrations (c.f. Table 8.1). At higher temperatures $\varepsilon'(T)$ tends to the value of pure BD. This is not surprising since methanol evaporates at 65°C.

From Fig. 8.5, below 65°C the conductivity of biodiesel increases with higher methanol content.

8.6 Concluding Remarks

Measurement data of flash point as a function of methanol content show clearly that small increases of alcohol concentration originate a steep decline in FP. This result is important for production, distribution and utilization of BD. In particular, excess of methanol must be eliminated with an adequate purification process.

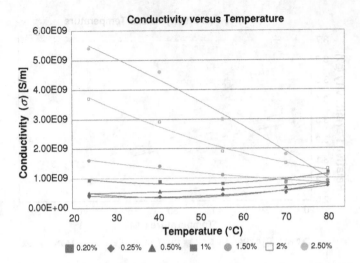

Fig. 8.5 Conductivity as a function of temperature for different values of methanol content in biodiesel

From the strong correlation between MC and FP, it is possible to use FP measurements for the preliminary verification (prior to certification) of the compliance with the methanol content requirements of BD standards. The importance of this results lies in the fact that FP measurements are much simpler and use less expensive equipment than the procedure for determination of methanol content recommended by standards (chromatographic analysis) [6].

The linear fit of permittivity measurements as a function of temperature makes possible to determine accurately MC in biodiesel within a wide range of concentrations, including the maximum value set by the standards (0.2%). Methanol content may be obtained from the value of $\varepsilon'(T)$ extrapolated to 0°C, that is, the parameter a in (8.3), given in Table 8.1.

Measurements of biodiesel conductivity at room temperature (25°C) may be used for a preliminary check of methanol concentrations above 0.5%, as shown in Fig. 8.5. This is very convenient since the equipment for conductivity measurement at room temperature is widely available and easier to use than systems for permittivity measurements as function of temperature, particularly in industrial applications [7].

Acknowledgments Figures 8.1 to 8.5 and most of the text have been reproduced with permission of Nova Science Publishers, Inc. from Ref. [7].

References

1. EN 14214 (2009) Automotive fuels—fatty acid methyl esters (FAME) for diesel engines. Requirements and test methods
2. ASTM D 6751 (2009) standard specification for biodiesel fuel blend stock (B100) for middle distillate fuels. ASTM International

3. ASTM D 93 (2002) Standard test methods for flash point by Pensky-Martens closed cup tester. ASTM International
4. González Prieto LE, Sorichetti PA, Romano SD (2008) Electric properties of biodiesel in the range from 20 Hz to 20 MHz—comparison with fossil diesel fuel. Int J Hydrogen Energy 33:3531–3537
5. Sorichetti PA, Romano SD (2005) Physico-chemical and electrical properties for the production and characterization of biodiesel. Phy Chem Liq 43(1):37–48
6. EN 14110 (2003) Fat and oil derivatives—fatty acid methyl esters (FAME). Determination of methanol content
7. Romano SD, Sorichetti PA (2009) Correlations between electrical properties and flash point with methanol content in biodiesel. Chem Phys Res J 3(2/3):259–268

Index